D0983731

SCIENCE IN CANADA

SCIENCE IN CANADA

Selections from the Speeches of E. W. R. STEACIE

Edited by J. D. BABBITT

UNIVERSITY OF TORONTO PRESS

WILLIAM MADISON RANDALL LIBRARY UNC AT WILMINGTON

© UNIVERSITY OF TORONTO PRESS 1965
Reprinted 1968 in the U.S.A.

SBN 8020 1363 5

Q127
.C2
.S7

"He was an outstanding example of a type that has always specially attracted me: the specialist who has a wide outlook, broad knowledge, and warm enthusiasms outside his own subject as well as in it, and, more particularly, a man he was, whose mind has been trained in the splendid discipline of a science, but whose heart and eyes take also delight in the triumphs of art, in the history of man, in the beauties of nature. Such a man is about the best thing that our modern civilization can produce."

SIR GEORGE TREVELYAN, IN SPEAKING OF A FRIEND

118144

Acknowledgment

IN PREPARING these selections from the speeches of Dr. Steacie, I am indebted to many people for advice and assistance. In particular: to Mrs. Steacie for her kind interest and her many sympathetic comments; to Mel Thistle and to Dr. C. J. Mackenzie for reviewing the text and for suggesting numerous improvements; and, above all, to Miss Ethel Wheaton, the faithful and efficient secretary to two Presidents of the National Research Council, who made the manuscripts of these speeches available to me, helped with the preparation of the text and, in many discussions, added to my understanding of Dr. Steacie's views.

The publication of this volume has been made possible by the generosity of the Canadian universities who, in recognition of the great role played by Dr. Steacie in the development of academic science in Canada, responded readily to an appeal from the trustees of the E. W. R. Steacie Memorial Fund for contributions to a grant in aid of publication.

J.D.B.

Contents

SCIENCE IN CANADA

Introduction

AT HIS DEATH in August 1962, Dr. E. W. R. Steacie was the accepted leader of Canadian science, a distinction that he had attained not alone because of his official position as President of the National Research Council, but also because of his outstanding research accomplishments, his sure feel for scientific values, and his strong personality. In Dr. Steacie, official position and personal ability were happily matched; it seemed almost that the wide responsibilities of the National Research Council had been designed with such a man in view. Although at his death his reputation and his influence extended well beyond the Canadian scene, it is for the inspiration and direction that he gave to science in Canada that Dr. Steacie will best be remembered.

It is to be regretted that during his life Dr. Steacie never took the trouble to develop his views on the organization and administration of science into a comprehensive study. It would never have occurred to him to do this. He was absorbed with his research and the immediate problems of his position, and, because he reacted always from fundamentals, he thought of his views as obvious and not requiring elaboration. He wrote little not bearing directly on his researches and, if he had not been forced by his position to give numerous addresses, he would have left no record at all of his ideas on the broader aspects of science.

Even before his death I had urged Dr. Steacie on several occasions to allow me to collect his speeches so that they would be widely available. I felt that Canadian scientists—including the staff of the National Research Council—were insufficiently aware of his ideas on the promotion of science in Canada and, more particularly, on the administration of government research laboratories. My opinion of the importance of spreading his views had been based on one or two individual speeches and not until I was in possession of the complete collection did I recognize their full value and realize how completely they portrayed his thought. The most striking characteristic of Dr. Steacie's writing is its pungent, almost aphoristic nature, but because he was always concerned with the essential issue, the speeches in their totality present a composite picture of the philosophy and style of thought that distinguished his approach to science and to the organization of scientific activities.

Dr. Steacie was, *au fond*, the practical experimenter, a scientist who saw clearly the problems of his field and systematically set himself to solve them. In everything he did he was personally well organized and it was always apparent that he was on top of his job. He was a man of great intellectual power, a man with a clear and incisive mind who moved resolutely and quickly towards practical solutions. He was little given to introspective theorizing but was, essentially, the man of action, a leader who in another age might have made a brilliant military commander or the great governor of a colony. His intellect and scholarship led him to chemistry, but his natural leadership and his grasp of essentials took him to the top position in science in Canada and to an international reputation as a scientific statesman.

But neither his scientific scholarship nor his ability as a

leader would alone have sufficed to give to Dr. Steacie his special significance. There have been many scientists who have made comparable contributions to chemistry and there have been scientific statesmen who have equally dominated scientific developments in their own countries. What gives to Dr. Steacie his unique quality and to his speeches their special flavour is the combination of the attitude of the academic scholar with the pragmatic approach of the man of action who knew intuitively what needed to be done and who moved resolutely to do it. His partiality towards the academic life did not lead him to the seclusion of the ivory tower nor, on the other hand, did his natural bent for leadership delude him into accepting authority as an end in itself; yet he understood fully the outlook and philosophy of the academic and was more efficient and more practical than any single-minded administrator. This comes through clearly in his speeches: he could deliver a Presidential Address to the Royal Society of Canada that, in scholarly insight, equalled the best efforts of the humanists in that multi-cultural society and yet in his comprehension of the intricacies of government administration he was unsurpassed by any Treasury official. He was, accordingly, trusted by the academic scientists and respected by the mandarins of the government service.

The speeches which form the basis of this collection were given over a ten-year period during which Dr. Steacie was President of the National Research Council of Canada. They show quite clearly the change of interest and the enlargement of scope of his thoughts during this period. Dr. Steacie came to the Research Council from McGill University in 1939 to be Director of the Division of Chemistry. In 1950 he became Vice-President and in 1952 President of the Council. Although as Director and as Vice-President Dr. Steacie had been exposed to a certain measure of official responsibility, his interest remained predominantly the direction of research; only when he became President did the executive and policy-making functions of his position begin to displace the experimental and theoretical problems of the laboratory as his major preoccupation. The development of these interests is reflected in the subject matter of his speeches; although arranged by subject, the speeches conform, none the less, to a rough chronological development so that the progression of Dr. Steacie's thought is evident. In the early speeches he discusses the concerns of the academic—the place of science in education and the compatibility of science with the humanities. These are but the broadened interests of the university research worker and, although they never ceased to interest him as being fundamental to a proper determination of the place of science in national life, the demands of his position forced him rapidly to apply his insight first to the practical problem of reconciling scientific research with government administration and later to the political problem of the relation of science to government.

Although the themes of the speeches vary, there is throughout the same fundamental approach and the same orientation of thought; whether he is insisting that science in the university must be an education and not a training, or discussing the relation of a government research institute to a centralized civil service, Dr. Steacie's views derive straight from his conception of scientific research as a creative activity comprising an essential element of philosophical culture. To him science was a scholarly pursuit and the scientist a creative individual. He could accept no image of science that did not leave to the scientist his independent initiative. From these beliefs can be traced those strong oppositions which run through the speeches: his aversion to all attempts to plan, organize, or co-ordinate science, his objection to professionalism, and his abhorrence of secrecy. To him these concepts struck at the very roots of a vital science and were to be tolerated only as recognized evils.

In reading these extracts it is essential to understand the

position of the National Research Council in Canadian science and, even more important, to appreciate Dr. Steacie's concept of this position and his interpretation of his own responsibilities as President. Because of its distinctive characteristic of combining in one organization an honorary advisory council, the executive body for large government laboratories, and the committee responsible for the government support of research in the universities, the National Research Council occupies in Canada a position that is almost without parallel in any other country. To Dr. Steacie, his duties as the chief executive officer of the largest laboratory complex in Canada was one of the less pressing responsibilities of his position; the decentralized control of the laboratories which gave the Directors of the Research Divisions almost complete responsibility for research programmes, together with his stipulation that the administration was to be the servant and not the master of the laboratories, left the President free to concentrate on his responsibilities for broad policy and for relations with the Council and with government. Although he was proud of the reputation that the laboratories attained as an outstanding government research institute and although he realized the great stimulus that the achievements of the National Research Council laboratories gave to research in Canada, it was the broad responsibilities of the Council which absorbed his attention as President. Dr. Steacie realized the ambivalent nature of the National Research Council and understood well its many responsibilities. It was, as he said, at one and the same time, a Government laboratory with duties to maintain standards and to certify instruments, an industrial research institute, a grant-giving foundation, and finally a body exercising many of the responsibilities of a national academy. He was vividly aware that it was these multiple functions—particularly the combination of active research scientists with an independent advisory council—that gave to the Research Council its strength and vitality and which put it in a position to give strong leadership to Canadian science.

Dr. Steacie was particularly jealous of the Council's position as an independent body of scientists responsible for many of the functions of a national academy. It was through these academy responsibilities that he viewed the Council as exerting its major influence as the general policy-making body for science in Canada. As the speeches show, he had no patience with the view that a policy-making body such as the Council should attempt to organize, or even, in any formal sense, to co-ordinate science. Like Polanyi,[1] he believed in the "spontaneous coordination of independent initiatives" and the exercise of those informal mechanisms which traditionally have been used by academies—the scientific meeting and the expert committee—was as far as he would ever allow Council to go in organizing Canadian science. His clear exposition of the dilemma faced by a grant-giving body when tackling the problem of weak areas in science shows how sensitive he was to the use of government grants to direct research in the universities.

In his discussion of the teaching of science in the universities, Dr. Steacie singled out two problems: the substitution of training for education, with the attendant danger of professionalism, and the antagonism between the humanists and the scientists. On both of these issues Dr. Steacie took a characteristically strong position based on fundamental considerations. This was not an isolated attitude, but stemmed directly from his innate belief in the individual and from his concept of science as a creative activity demanding an imagination as vivid and a scholarship as exacting as the humanities. The designation of scientists as "only glorified technicians"[2] obviously irritated him and he could not tolerate the pretension that the humanities were without any stain of narrowness

[1]Michael Polanyi, *Minerva,* I (1962), p. 54.
[2]*Report of the Royal Commission on National Development in the Arts, Letters and Sciences* (Ottawa: King's Printer, 1951), p. 138.

or any trace of utility. He would grant to the humanists neither their claim to be the sole source of philosophical enlightenment nor their right to attribute to science all the problems of professionalism. But unlike other scientists, Dr. Steacie placed great value on the humanities and the social sciences; he recognized that to them we are mainly indebted for the academic atmosphere of the university which has proved so congenial and so fertile for scientific research and he took issue only with the attitude of superiority sometimes found in academic humanists of negligible accomplishments. Dr. Steacie was very conscious of the dangers of an education of the Soviet type, directed solely to the economic benefits of technology, but he believed that it was essential that the traditional values of the humanists should be refreshed and invigorated by the new philosophical outlook of science. In his view a new humanism is required today, a humanism which will not ignore the great advances of the sciences but which, on the contrary, will place them in their proper perspective along with the traditional values of the humanities. As the speeches show, Dr. Steacie was a keen student of the history of technology and he follows Sarton in believing that the most promising way to break down the division between the "two cultures" is through the historical study of science and technology.[3] He points out on several occasions that, unless some method is found to incorporate a knowledge of scientific values into Arts courses, the increasing importance of technology will drive more and more of the ablest students into science and engineering with perhaps disastrous results for scholarship in Canada.

In his Presidential Address to the Royal Society of Canada, Dr. Steacie applied himself to a problem that became of increasing concern to him, the problem of advice to governments and the way in which the power and influence of national academies have largely been replaced by official

[3]George Sarton, *The History of Science and the New Humanism* (New York: H. Holt and Co., 1931).

advisers. Dr. Steacie's advice to the Royal Society was to avoid "vague high-sounding objectives" and to concentrate on "real problems open to attack." Because he placed such great importance on the functions that the National Research Council carried out in lieu of a national academy these remarks are of more than ordinary interest. He was addressing the Royal Society but he undoubtedly was thinking of his own experience as Chairman of a Council with advisory functions not dissimilar from those of an academy. With his instinctive ability to distinguish the spurious from the genuine, Dr. Steacie here struck truly the balance between the sterile and the useful advisory committee. He is, in essence, repeating the view of K. C. Wheare that committees set up to advise on wide general areas "are of relatively little value" in comparison with the committee of experts set up to advise on a particular problem in which they have special knowledge.[4] Dr. Steacie had always a great scepticism of committees (as of teams) but he realized that in the Honorary Advisory Council of the National Research Council he had a committee which, as the executive body for the laboratories and in its responsibility as a grant-giving foundation, dealt with "real problems open to attack." These were, he realized, functions proper to a committee, functions, in fact, requiring the composite views and the joint responsibility of a committee and, during his term as President, Dr. Steacie so guided and led the Council in its deliberations that it gained a merited reputation as one of the most effective bodies of its kind in the world. Since his death, however, the Council has been criticized because it did not attempt to develop a "cohesive" plan for science in Canada or to advise the government on "broad national policy."[5] If this criticism is justified, it is Dr. Steacie who must take the responsibility

[4]K. C. Wheare, *Government by Committee* (Oxford, 1955), p. 62.
[5]*Report of the Royal Commission on Government Organization* IV (Ottawa: Queen's Printer, 1963).

and not the Council. As these extracts make abundantly clear, he was implacably opposed to any attempt to formulate a broad general plan for science. But this is not to say that he himself did not have a firm conviction of what constituted sound policy or did not know very clearly where science in Canada should go.

From the beginning of his stay at the National Research Council, Dr. Steacie had concerned himself with the problem of the administration of government research laboratories: first as the research scientist to whom administration was an evil to be avoided at all costs and latterly as the executive officer who had to ensure that his own administration was free from all the faults he had condemned as a scientist. As with everything he tackled, Steacie's approach to administration was from first principles; he was concerned always with ends and never with means and to him the end possessed value only in scientific terms, not as an administrative organization nor as an operational structure. His articles of faith were that scientific establishments should be run by scientists, that the organization must be made to fit the man, and never the man the organization, and above all that administrative considerations must never be allowed to dominate. As he stated to the Parliamentary Committee on Research, in his own responsibilities he practised these principles to the extent that he strove always "to make sure that administration can never issue any instructions to scientists in connection with any technical subject whatever."

Holding these strong views on administration, Dr. Steacie was always alert to any attempt to undermine the statutory position of the Council as an independent agency and he was not above taking action contrary to Civil Service regulations simply to assert his independence. When in 1958 the Civil Service Commission published their report advocating that the National Research Council (along with certain other "exempt agencies") should be brought within the Civil

Service Act in order to form one vast uniform Civil Service, he responded immediately to the challenge and characteristically applied all his energy and his ability to defend the position of the National Research Council.[6] The culmination of his defence was reached in the talk which he prepared for presentation at the Institute of Public Administration. Before this talk could be delivered, however, he was assured that the independence of the National Research Council would not be touched and, with his objective attained, he prudently refrained from delivering an address which might further have alienated that portion of public opinion which was already inclined to think that he placed the National Research Council on too high a pedestal. The address does present so forcefully, however, all the points that Dr. Steacie held to be essential for the effective operation of a government research institute that it is reproduced here in its entirety.

Although Dr. Steacie has been criticized in some circles as being excessively academic in outlook and as insisting too strongly that a leaven of fundamental science is essential to invigorate the research of government laboratories, his sympathetic analysis of the problems of industrial research in Canada shows how clearly he understood the shortcomings of Canadian science and how fully he appreciated their implications. By predilection and by training, Dr. Steacie was undoubtedly attracted to academic research, but it was equally true that when, as head of the National Research Council laboratories, he emphasized the necessity to avoid an excessive concentration on *ad hoc* projects and when, as President of the Council, he gave precedence to university research, he did so in the firm conviction that in the long run this was the wise way for the National Research Council to fulfil its responsibility to industry. He could not conceive of an effective industrial research complex without strong

[6]Civil Service Commission of Canada, *Personnel Administration in the Public Service* (Ottawa: Queen's Printer, 1959).

universities, and he saw the government laboratory as a support and stimulation for research in industry itself, but with a complementary rather than a competitive function. The contribution of the National Research Council laboratories was to assist industry with long-range research, with solutions to their more pressing problems, and with information and ancilliary services not otherwise available in Canada. Above all, Dr. Steacie knew instinctively that the greatest contribution that the Research Council could make to industrial research in Canada was to build up the graduate schools in the universities and to ensure that there was in Canada the scientific climate and the technical outlook that are today the distinguishing features of advanced industrial societies. In this, the verdict must be that he was successful.

Dr. Steacie owes his dominant position in Canadian science as well as his reputation on the international scene to the powerful combination of a great research scientist and an outstanding administrator. He was a man who saw all issues in fundamental terms and he expressed his views trenchantly and with great vigour. He was implacably opposed to the concept of the scientist as the Organization Man; he valued in the scientist the qualities of individuality and independence; the committee and the team were to him the personification of the woolly thinking he abhorred; he resisted all efforts at the bureaucratization of science and insisted throughout that the man was everything and the project nothing. In these days, when so often the emphasis is all on co-ordination and priorities, it is refreshing to read again a scientist who not only expresses so forthrightly the opposite view, but who in his own life's achievement was able, over a decade, to make this view the science policy of his own country.

Ottawa J. D. BABBITT
April, 1964

SCIENCE AND THE UNIVERSITY

Science and Education

ST. GEORGE'S SCHOOL
MONTREAL, JUNE 12, 1959

AS SCIENCE has become more important in our lives there has been a great deal of interest in the relation between science and education. Unfortunately, however, far too many people seem to have taken extreme stands, the arguments ranging all the way from the idea that it is a waste of time and effort to do anything but train everyone to be a scientist or engineer, to the opposite extreme that all scientists and engineers are of necessity uncouth, ignorant, and essentially uneducated. I hasten to say that I am going to take a middle position.

There are really two quite separate aspects of the problem. First, what is the place of science in a general education? In other words, how much science should be taught to an intelligent person who is not going to become a scientist? The second question is to what extent should a scientist be educated in other things? The ideal situation, of course, would be to teach everybody everything, but life is too short. Essentially, therefore, what we have to decide is what compromise to make, and the virtues of breadth *versus* specialization.

The Place of the School

First, there is the question of how the school comes into all this, and here one can obtain some guidance from history. There has always been a tendency for the university curriculum to move with the times. The original universities, 700 years ago or so, taught people to read and write and not much more. (Their education perhaps looked more dignified because it was Latin that they were taught to read and write.) At a time when almost no one *could* read or write this appropriately constituted higher education. As general education has increased in depth and become more widespread over the past seven centuries, the universities have continually sloughed off the more routine items to the schools. A junior clerk now requires much more education than was possessed by many university graduates a few centuries ago. In this connection the word "continually" should be used with some reservations because there have been periods of decline and decay in the universities. It is essential that this process of sloughing off continue if the universities are really to deal with *higher* education. With the changing demands of society for educated people it is vital that the curricula of both the schools and the universities move forward.

The Importance of Science

As science has become more important to society as the basis of technology, of defence, and perhaps of survival or the reverse, it might have been expected that the sharing of education between the school and the university would have been modified so that more and more science and mathematics were included in the schools. This would mean that the universities could raise their sights in science and start their teaching at a more advanced level. At the same time an increase in the teaching of science in schools would ensure that the general public of the coming generation would have some appreciation of the objectives and the method of operation of science.

In reality, however, a great deal of the worry over present-day education is due to the fact that exactly the reverse is happening. In many cases, as science has become more important to society the schools have given it less attention, and the university curriculum is obliged to move backward rather than forward.

Science and a Broad Education

In discussing the foundations of a general education it is impossible to avoid some reference to the unfortunate debate which has been going on as to the relative merits of the humanities and of science as part of the process of creating a "broadly" educated man. There is often the suggestion that the humanities are intrinsically broader, and that a narrow education in, say, the classics will produce a broader man than an equally narrow education in biology. It is perhaps worth pointing out that this argument that one part of knowledge is superior to another is of a relatively recent date.

The revival of classical learning at the time of the Renaissance was not an attempt to push one part of learning rather than another. It was merely the attempt to go back from a state of barbarism to a much higher level of learning which had prevailed ten or more centuries previously, and to rediscover the knowledge and learning of classical times.

This was soon followed by a revival of interest in learning in all fields and the new "science" was vigorously pursued in the sixteenth and seventeenth century by poets, philosophers, theologians, and classical scholars. By the eighteenth century science had become a fashionable pursuit of educated men, but the universities were in an almost total decline as far as real scholarship was concerned in any field. After the revival of the universities in the nineteenth century, science became a part of university education in a real way, but it was about this time that the idea of the superiority of the classics began to be a rallying point for conservatism.

There were, of course, good reasons for the type of education offered by the universities in earlier days. The *Oxford Dictionary* (compiled mainly from 1900 to 1920) defines a liberal education as education fit for a gentleman. Ashby remarks that this is still an acceptable definition: it is the idea of a gentleman which has changed. To quote Sir Eric Ashby,

> A century ago, when Britain awoke to the need for technological education, a gentleman belonged to what was called the leisured class. The occupations of his leisure did not require any knowledge of science or technology. Modern gentlemen do not belong to the leisured class—and more and more of them are finding that their business requires expert knowledge. Even members of the House of Lords are called upon to make decisions about radioactive fall-out, overheating during supersonic flight, and the strontium content of bones. Even such a gentlemanly subject as the state of the river Thames cannot be understood without some knowledge of oxidation and reduction, detergents, and the biochemistry of sewage.[1]

This raises the very important question of the status of science in a broad education. The change in man's civilization, outlook, and knowledge in the last 300 years constitutes a revolution as great as that of the Golden Age of Greece. Can one ignore all this and still have sufficient breadth of education to decide where society is heading? The major new factor today is man's ability to exercise control over his environment. It is difficult to see how a man can express contempt for his environment and all knowledge of it and still claim to be educated. In short can you deal with the "whole man" while neglecting his environment altogether? These arguments have particular force in Canada because, as an animal, man is here very far north of his "range" and can live here only because

[1]Technology and the Academics: An Essay on Universities and Scientific Revolution (London: Macmillan, 1958), p. 81.

he does possess control over his environment. It is necessary in Canada to have some appreciation of the facts of life.

These questions pose a serious problem for education. Somehow it is necessary to give future leaders of society some general idea of the aims and methods of science, the frequent lack of which is a great complication in present society. Some understanding of science is necessary not only in order to make appropriate decisions for the welfare of science, but also as self-protection against scientists who might be putting something over. When our whole technology is based on science it is certainly as important that it be generally understood as it is to have economics generally understood. The problem, however, is not easy, and is not to be solved by cramming elementary chemistry or physics down the throats of first-year Arts students. What are necessary are not the efforts of the popularizers of science who write semi-fictional articles about its wonders, or try to explain the whole of nuclear physics in one easy lesson. The effort must rather be to give some idea of what science is and how it works, of its philosophy and methods. If possible, the ideal situation would be to try to give everyone a knowledge of one science in some depth, rather than the more usual method of trying to give a smattering of everything. Another method of attack is through history. There has been a highly desirable trend in the study of history away from too many wars and dates and towards more social history. It is peculiar in this connection how little attention has been paid to the history of science and technology which could well give greater insight into the real workings of science than more formal courses of the Physics I type. Certainly the school as well as the university could do much along these lines. . . .

Science and the University

THE UNIVERSITY OF BRITISH COLUMBIA
MAY 21, 1957

THE POSITION of the university in the field of science is a subject which is much under discussion these days, and is at the same time widely misunderstood. A major feature of our times is the rapid expansion of technological organizations and the very rapid rate of technological innovation. Because modern technology is based on science the welfare of the two is obviously related. At the same time the fact that technology is based on science has produced in the popular mind a considerable confusion between the two and, in particular, an inability to understand the difference in the outlook and aims of science and technology. The picture is also complicated by the fact that there are those who feel that the rapid rise of technology is incompatible with the proper appreciation of the humanities and of the finer things in life. They seem to be uncertain, however, whether to blame science or technology.

It is worth going back for a moment to consider the development of science and of the practical arts, or technology. From the earliest times until about three hundred years ago the industrial arts, almost without exception, advanced solely, and very slowly, by a purely empirical process of trial and error. Tradition and the conservatism of the crafts tended to carry methods on unchanged from generation to generation. In some important fields, such as roads and water supplies, there were periods of as much as one or two thousand years without appreciable advance in methods. (Municipal affairs seem to have proceeded slowly in those days too.) Carpenters' tools in the Middle Ages were almost indistinguishable from those in use two thousand years earlier.

During all this period, up to, say, three or four hundred years ago, the "explanation" of natural phenomena had been based mainly on a combination of superstition and an appeal to tradition. There are, of course, some fields of human endeavour which are still in an analogous situation. With the development of the so-called scientific method, which is essentially merely an attempt to be objective and to ignore tradition and appeals to authority, curiosity about natural phenomena became widespread. Probably the best summary of the scientific method is the motto of the Royal Society of London, which was founded three hundred years ago. It is "nullius in verba," and there is no doubt that a refusal to take words too seriously is a major reason for a certain distrust of scientists by those whose main tools are words.

The development of the scientific method, which was essentially the birth of science itself in any real form, led to attempts to explain things which were well established in the industrial arts; in other words, science began to have an application to technology, and the pace of technological innovation began to increase rapidly. It is rather interesting to note that as long as the industrial arts advanced by empirical means the Western World was never very good at the game. We are apt to overlook this and to feel that we have always been far ahead of everyone else in the practical things of life. In fact, however, until quite recently Chinese technology was in the main far ahead of that in the West. Many examples could be cited in which the West lagged behind the East by as much as a thousand years. A few outstanding cases are: the wheelbarrow, by 10 centuries; the cross-bow, by 13 centuries; proper harness for horses, by 6–8 centuries; canal locks, by 17 centuries; and printing, by 6 centuries. It was only when technology began to be based on science that the West caught up with the East and, in recent years, of course, far outdistanced it.

The development of technology based on science has continued to accelerate with the introduction of industrial and military research. By 1920 natural knowledge had advanced to a stage where the large-scale application of science to industry became profitable. As a result the stock of the scientist rose at the expense of that of the inventor. The major change produced by this in the last 30 years has been the rise of the industrial research laboratory and the advent on a significant scale of the professional research worker.

It is rather difficult for any one under 40 to realize the change which has occurred in the last 35 years. In the 1920's research was a vague and rare thing done by university professors for reasons best known to themselves. At all events it was popularly regarded as a highly impractical pursuit. Ph.D.'s were rare birds, so rare in fact that any one with the title "Doctor" was apt to find himself receiving unexpected medical confidences. It is not surprising that the transition to the professional research worker did not occur without some difficulties and dislocations. In fact, science has gone through changes exactly analogous to those of hockey and football. The change from amateur to "semi-pro" to professional has created some problems and, while the standards of play have improved, some intangible things have been lost. Also, popular interest has risen to such a degree that research has to some extent become a "spectator-sport."

The interlocking of science and technology has led to considerable confusion about the aims of science. It is, of course, the function of science to enquire into the workings of nature, and the application of such knowledge has become the mainstay of technology. The motives of science and technology are thus distinctly different although in many cases their methods may be very similar. Science is thus in a dual position as part of a humanistic education (after all it *is* a branch of philosophy), and as the basis of technological development. One danger of the importance of science to technology is that science in its own right as a branch of knowledge is apt to be overlooked or minimized. It is, in fact, the ignoring of science in its own right which has been responsible for drumming up a purely bogus clash in outlook between the scientist and the humanist.

There appears to be no question that from all points of view it *is* desirable to attempt to understand more about the working of nature, and the pursuit of science for its own sake is definitely a fundamental part of the responsibilities of a university. There seems also to be a clear obligation to society to develop the knowledge which ultimately underlies technology, and to educate the people who will later contribute to technological development. It should be realized that universities were not always the main location of scientific work. Three hundred years ago, when workers were amateur and equipment was simple, most work was done in private laboratories. The universities were by no means keen on letting science creep in the door, and it is only within the last century that it has been regarded as a reasonably respectable university subject.

Also it is important to remember that universities have not always maintained their position as centres of creative scholarship. The universities, in fact, declined in the seventeenth and eighteenth centuries into organizations of social conservatism which gave little stimulus to creative work. As a result outstanding scholars of the day, such as Bacon, Descartes, and

Boyle had almost no connection with a university. With the re-vitalization of the universities in all fields in the nineteenth century, science developed rapidly, and until about 30 years ago almost all work in pure science was done in universities. This is certainly the proper place for it. The university atmosphere is undoubtedly appropriate for the search for knowledge for its own sake, just as the industrial atmosphere is proper for work on the application of science to technology.

The rise of industrial research in the last 30 years has, however, brought about some changes. A large amount of research, including an appreciable amount of fundamental research, is now done in industry and in research institutes rather than in universities. This work is of the greatest importance to the development of technology, but the situation is dangerous if it proceeds to the point where the university is no longer the main home of pure science. The chief reason why the university is the ideal place for scientific work (the social sciences included) is that the work is uncommitted. The university man is free to proceed in any direction which he sees fit, and should not be in any way influenced by practical considerations. The university is, in fact, virtually the only place where science can be pursued for its own sake.

In recent years there has been a very sharp rise in the cost of equipment and facilities which has led to financial pressure on the universities and has caused them in many cases to accept "sponsored" research projects, that is, projects with a technological motive. Such support can be very helpful provided that it is for problems chosen by the investigator. Often, however, the support is for specified projects, and the effect of such work can be most unfortunate. It leads to lack of freedom to follow whatever path the worker may see fit, and to outside planning of university research. No matter how fundamental such work may be, there is still an element of ulterior motive which is regrettable. It is most desirable that the universities be put into a financial position where such outside pressures can be resisted, and that the university remain the centre of the free and unrestricted pursuit of science for its own sake.

This is essential not only from the point of view of the welfare of science, but also, I think, equally important in creating the only atmosphere in which proper scientific and engineering education can be carried out. In the day of a very heavy demand for scientifically trained people, it is also important that the dual role of the university to teach and to advance knowledge should not be overlooked. The last decade has produced a most striking advance in the quality and quantity of scientific research in Canadian universities. It would be most unfortunate if in the future we produced more scientists at the expense of the quality of university education and of the university's part in the advancement of science.

The Task of the University Today: Science

ROYAL SOCIETY OF CANADA
JUNE, 1960

IF WE ARE TO DISCUSS the educational responsibility of the universities, the first and the most important question is to whom should they be responsible? There are various possibilities, the first being the general public. I doubt if the universities can delegate their responsibility or expect much guidance from the public, although they should certainly feel a responsibility to the public to do the best they can for them. The university must lead rather than follow, however, and I can think of nothing worse than for the university to follow the direction of public opinion as it drifts aimlessly about under the influence of periodic pressures.

A second possibility is responsibility to the government or governments. In most cases there *is* financial responsibility. However, nothing could be more detrimental to the real purpose of the university than to follow blindly the wishes of the government of the day. Certainly it is not the function of the university to turn whole-heartedly to the production of fascists in a country which heads in that direction, or of communists if it heads the other way. It may, of course, be very difficult to resist such pressures, but it is noteworthy that when a blow-up occurs in a totalitarian state it is usually precipitated by students, and that a dictatorship is often followed by the closing of universities. Certainly a university may often have the duty of opposing the government of the day. In a less extreme example one would certainly not expect economics to be slanted one way or the other as governments change, although occasionally even this type of pressure has been exerted. On the other side of the picture, there are many indications that parliamentary institutions do not work well in countries which do not have a background of academic and educational freedom.

Again, there is the responsibility to industry. Certainly, because most graduates will take their place in industry, there *is* a responsibility not to ignore the needs of industry. A direct responsibility to produce the kind of graduate industry wants would, however, be a degradation of the university to the level of a vocational school.

It is frequently stated that the real responsibility is to society. This, however, is a rather amorphous term, and means nothing without further specification.

The only real responsibility of the universities is to remain true to their purpose and traditions, and to decide for themselves what will best fit the needs of government, of industry, and of society over the long haul. Surely the universities should regard themselves as the most enlightened and objective fraction of society—even if, in some cases, they are not!

This is not to suggest that universities should be ultra-conservative, should refuse to move with the times, or should ignore the needs and demands of society. I agree strongly with Ashby when he points out that the ability to move with the times and to adjust to the changes in society has been the major characteristic of universities over the last 800 years, and that it is this ability to "roll with the punch" which has enabled the universities to survive almost unchanged in form and in basic principles over eight centuries. The important thing is that in such adjustments the university remain true to its own basic principles.

In this connection one of the more disturbing slogans of the day is "Education is everybody's business." That everyone should support education is, of course, undeniable; the theory that everyone should regard himself as an expert in education is, however, a most unfortunate one. The last few years have seen many conferences on higher education, often attended by a peculiar collection of people whose only common bond has been ignorance of the subject under discussion. Such conferences have often resulted in resolutions of a high degree of irresponsibility, but fortunately of a high degree of ineffectiveness as well. Universities, of course, have always had patrons and patrons are always difficult, but today's multiplicity of them is a new and perhaps particularly difficult experience.

One of the problems which the universities have to face in dealing with the demands of society is that of training as opposed to education, a problem often coupled with criticisms of specialization. One famous Canadian document says that scientists are often "only glorified technicians, lacking any broad understanding of the field in which they labour."[1] The arguments for this point of view are dubious, and I reject them emphatically, but there is an aspect of specialization which needs to be watched, and that is the one which verges on training. Training for the tasks of life is not necessarily incompatible with a real education—always provided that such training is based on the acquisition of knowledge of fundamentals. If, however, the curriculum is overloaded with practical odds and ends the process becomes purely training and ceases to be real education. There is no question that this is a serious problem and it is becoming more serious. In parti-

[1]*Report of the Royal Commission on National Development in the Arts, Letters and Sciences* (Ottawa: King's Printer, 1951), p. 138.

cular, we must guard against university courses which lead to mere professional qualification. Engineering has been a source of worry from this point of view, but there are signs of an encouraging trend back to principles. I must confess also that I object strongly to the recent interest in "manpower." The tendency to regard manpower as a commodity bought, sold, produced, and consumed is objectionable when applied to the end-product of an institution of higher learning. It is, moreover, a dangerous concept in that it implies the unimportance of quality and the importance of mere numbers. It is also dangerous from a university point of view in that shortages can become surpluses with the greatest of ease, given a slight financial recession.

The major problem in all this is professionalism, and the modern university can, I think, be justly criticized for permitting an excessive infiltration of courses aimed at a narrow phase of technology or of a profession. There is a curious attitude of superiority on the part of Arts faculties which overlooks the fact that most such courses are not in engineering or science. Technology is by no means confined to engineers and scientists; it is merely the sum total of what most of us do. Arts graduates are as much concerned with technology as engineering graduates, and a course in the economics of fish marketing is just as technological in content as one in fisheries engineering. In fact, courses in business administration are much more technological in content than courses in engineering, and courses in education are probably more narrowly professional than anything else that the university has to offer.

In a technological society, of course, the university can hardly remain aloof, but it is certainly open to criticism when it allows itself to be made the instrument of mere professional qualification. The important point is whether the student leaves an institution of higher education with a basic education in the principles of his subject, or with merely a professional qualification to take his place in industry or government with a minimum of difficulty and expense to his employer or to himself. This is the distinction between education and vocational training.

The difficulty, of course, is that there are many bogus professions as well as real ones, and there are many pressures, local and otherwise, on universities to give courses with a specific professional aim. It appears to be the desire of almost every working group today to convert themselves into a profession, to have the profession closed by legislation, and to get a specific university degree in their own subject. As far as subjects go, it is instructive to compare biophysics with cosmetic chemistry. Neither are professions, but there is a fundamental difference between them. An interdisciplinary course like biophysics is in no sense a narrower specialty than physics or biochemistry. The subject merely covers an area that crosses traditional borders, but is a fundamental area of knowledge. It thus qualifies in every way as a proper subject for university instruction. The details of such a course may be difficult to arrange, but the principles are clear. Cosmetic chemistry on the other hand is in no sense a real field of knowledge; if such a course is given (and it *is* given in some universities) it is merely a hodge-podge of odds and ends for the benefit of a specific industry.

It is difficult to draw the line between real and bogus professions, and, as usual, dictionary definitions are not helpful. Certainly there are real professions such as medicine and law where the protection of the public demands a closed profession, legally regulated. Even these, however, have their dangers for the university because the existence of an outside regulating body means that the university has, in principle, lost all real control of the content of courses. In fact, in both cases the problems have largely been avoided by informal action, but the dangers still exist.

There are, on the other side of the picture, many professions which are more or less bogus and which cause vigorous debate. I am going to keep carefully clear of library science, "education," and many others to avoid controversy, and just mention one for which I presumably am qualified—chemistry. There are those who would like to see chemistry equipped with all the trappings of a closed profession, including legislation, licensing, and outside control over university courses. Personally, I do not believe for a moment that chemistry should be regarded as a profession. There is certainly no need to protect the general public, and there is everything to be said for allowing any university to give the kind of course it wishes, and for anyone to practice chemistry whether formally trained or not. Rutherford obtained the Nobel Prize in chemistry without a degree in the subject, and Faraday, who had no degree at all, would certainly have been turned down by any licensing body. The point I want to make is that these licensing efforts constitute a real danger to university autonomy and to academic and personal freedom, and should certainly be closely watched in the future.

In connection with peculiar university courses I hope you will pardon a certain amount of levity if I give the result of browsing through the *Commonwealth Universities Yearbook* and *American Universities and Colleges*. There is a widely held, and unfair view that the United States leads the field in unusual degrees. This is quite untrue. The Commonwealth is far ahead. The two handbooks together list the titles of nearly a thousand different degrees. The Association of Universities of the British Commonwealth includes "Bachelor of Rural Science," "Master of Science in Glass" (which presumably leads to a Ph.D. in aspic), and "Bachelor of Expression." In the United States there is "Bachelor of Science in Group Work Administration," "Bachelor of Science in Practical Arts and Letters," or, because the B.Sc. has been preempted for so many things, the logical degree to end all degrees is "Bachelor of Science in Science."[2] I will tactfully refrain from citing Canadian degrees, except the most poetic, the B.Sc. (P.O.T.).[3] All this does emphasize how far the universities have travelled in 30 or 40 years from the time when there were a mere handful of degrees available. Incidentally, it is noticeable that when an Arts faculty wants to give a degree of which it is ashamed it always gives a B.Sc. with an appendage, rather than a B.A. This is gross discrimination against science!

Whatever may be said about trivial or vocationally slanted degrees, I should like to emphasize that I am not criticizing interdisciplinary courses, which are of increasing importance in covering the gaps created by hard and fast boundaries between fields of knowledge. Such courses are needed not only within faculties, but more are badly needed between faculties also, and one of the most effective developments of the last few years is the course in engineering physics. There are, also, possibilities for courses which bridge arts and science, and arts and engineering. In particular, many engineering graduates proceed directly to administration in large, technically-based industries. Are there not here possibilities for more courses which span economics and engineering, and for considerable technical content in courses in business administration?

Finally, of course, there is the perennial question of how to bridge the gap between science and the humanities. The need for people with a broad general education is as great, or greater, than it ever was; but the question is, who is broad? By refusing to take any interest in man's environment, there is no question that the humanists have surrendered their right to regard themselves as liberally educated and have left a gap which must be filled. The much talked about "whole man"

[2]*American Universities and Colleges* (Washington: American Council on Education, 1948), p. 1013.

[3]*Canadian Universities and Colleges* (Ottawa: National Conference of Canadian Universities, 1958), p. 57.

must have knowledge of his environment as well as of himself, and it seems to me that the time has come to recognize the need for the whole man and to stop arguing about which part of him is the top half. . . .

NATIONAL CONFERENCE OF CANADIAN UNIVERSITIES
OTTAWA, NOVEMBER 12, 1956

. . . As enrolment increases and classes get larger it is apparent that the real university function is being exercised more and more at the graduate level. In fact, at least in science, one may make a strong case for undergraduate training being merely a preparation for a real university education in the graduate school. This means that the graduate school of the university is becoming more and more important. The protection of the university position in graduate work is one of the most difficult aspects of the coming crisis. The problem is one of staff to a much greater extent than space. If the university can plan ahead, and secure staff when they can be found, the coming expansion can greatly strengthen the university. If, however, things are left to the last minute the university's graduate effort can be wrecked for a decade by the recruitment of second-rate staff. . . .

UNIVERSITY OF MANITOBA
MAY 19, 1954

. . . A second way in which universities are hampered is by way of advice from outside on the type of training the university should give. In the first place, such advice usually is an attempt to degrade the university to the level of a vocational school. In the second place, the only thing which saves most industries and laboratories from stagnation is the fact that the newly-hired graduate is much more up-to-date than his boss. Nothing worse could happen to an industry than to hire people trained exactly as it wished them to be. . . .

ST. FRANCIS XAVIER UNIVERSITY
AUGUST 29, 1957

. . . There seems to be no doubt that the university is the best place for pure science; the question is whether the universities can maintain their position in pure science. I would, therefore, like to review briefly the main points arising from the changes which have occurred in the last 30 or 40 years.

Professionalism

The change from the amateur to the professional has led to the rise of the professional society and the professional attitude. There is no doubt in the case of subjects like medicine and law of the necessity of this, but the development of a professional attitude towards research and creative thought is a different matter.

Size

A striking feature of the situation has been the expansion of research and of the so-called research team to enormous proportions. Size has inevitable disadvantages. Some of the more obvious are red tape, the necessity for liaison, information officers, the writing of endless trivial reports, the setting up of committees, and so forth. There seems no doubt that some of the very large organizations have reached, or passed, the point of diminishing returns.

During the war a frequent wise-crack was that if the war

went on much longer every scientist would be a liaison man between two others, all work would stop, and liaison would become perfect because nothing at all would be happening.

I remember on one occasion, when I needed to find out something in a hurry and rather doubted if anything at all had been done on the subject, I was surprised to find that we had 26 secret reports in our library. On reading them, however, it appeared that one single experiment had been done. However, reports had been written on the proposed experiment, interim reports, summarizing reports, and ultimately a final report had all been compiled. There seems to be no doubt that size has its drawbacks.

Size also leads to two very dangerous activities. These are planning and co-ordination. Planning means merely that someone in a high administrative post, with no knowledge of the subject, lays down the programme in detail. It is then taken over by the co-ordinator; he is relatively junior and knows nothing at all. In the ideal situation, under these circumstances everyone knows what everyone else is doing and thinking: precisely nil.

Another thing that goes with size is, of course, efficiency. An efficient organization is one in which the accounting department knows the exact cost of every useless administrative procedure which they themselves have initiated.

The universities are today in a very difficult position. Increasing competition from industry for people, increased costs of equipment, and so on raise the question whether the university can continue as the major location of scientific work, or whether industry will take over. It would be most unfortunate if the university lost its place, because it is the only place where work is uncommitted and can be done for its own sake, without commercial motives coming in. I must say that I am not as pessimistic as some. I realize the difficulties of the university financial position, and I realize that there must be improvement. However, the university still has some great advantages over the large industrial or military laboratory.

In the first place, the university laboratory is relatively small. Large organizations, well planned and co-ordinated, will drive ahead relentlessly to the obvious conclusion, but are most unlikely to produce original ideas. Teams are all very well, but neither a team nor a committee has ever been known to have an idea. Ideas come from people, and the universities should remain the major source of them. A second advantage of the university is that it is inefficient in the cost accountant's sense of the word. This is its greatest hope. There is nothing more antagonistic to original thought than business efficiency. In fact, as long as the universities can remain inefficient there is hope for the world. Universities also have maximum freedom of expression. Further, they have youth and its enthusiasm. Later the students will begin to worry about their responsibilities, salary, career, and pension.

All these factors indicate that the university has a real, vital, and continuing place in science. Youth and freedom make up for much in the way of equipment, but the university must not be asked to work at too great a disadvantage.

Some Problems of a Granting Body

CANADIAN FEDERATION OF BIOLOGICAL SOCIETIES
GUELPH, JUNE 1, 1961

. . . FROM THE POINT OF VIEW of academic freedom the increasing cost of research is a very important factor. In the early days of the present century relatively little equipment was necessary for most types of work. What was mainly required was an endowed chair which gave the occupant time to carry on investigations with an assistant or two. Endowments were often ample for the purpose, and there was no necessity of outside support (or interference). Today equipment is usually expensive. Further, individual investigations require much more elaborate techniques and take much longer. As a result it is not usually possible to make a significant contribution without a group of students. Each of them requires personal financial support and expensive equipment. It is highly unlikely that any university can afford all of this from its own unencumbered funds and continual outside assistance has thus become essential. The problem is how to obtain and accept such support and at the same time maintain the academic freedom which is essential for the welfare of the university and the success of the work.

There are two main types of outside financial support for university research on a year-to-year basis. A "grant-in-aid" is essentially a grant given to a university professor to assist him in carrying out work which he himself wants to do on a problem he has chosen himself. In theory at least this type of grant does not restrict the liberty of the recipient in any way. At the other extreme is the "contract" with a university research department when the university professor is hired to do a specific job specified by the purchaser. In principle this restricts his freedom totally. In between these two extremes lie grants for research on rather narrowly specified projects, for example, the professor is free to do anything he likes as long as it concerns the manufacture of glue. This type of grant is not quite as objectionable as the contract, but is more dangerous because it may give a false appearance of freedom of action.

Government support of university research in Canada has increased very rapidly over the last 30 years, especially since the end of the war, and there is no question that much of the research in universities could not continue if it were not for this support. There are two types of danger to the freedom of the individual worker as far as these grants are concerned. In the first place, the contract or grant for a specific investigation not chosen by the investigator seems to the writer to be dangerous, except as an occasional special case. The function of a university is not that of a consulting firm, and the purchase of services can be done only at the expense of the proper university function. Occasionally in engineering fields such contracts are justifiable, but in general they certainly do not encourage real research. In addition they have the highly objectionable quality of steering scientific work in a direction set by outside bodies. In the long run this type of grant is certain to encourage work with a narrow short-term objective rather than the uncommitted type of investigation for which the university is especially well fitted. There is also the very great danger in such specific grants that the granting body is likely to want to "co-ordinate" work of various workers. This ultimately can lead to disastrous attempts to "plan" pure science, and in the long run to degrade the university worker to the level of a technician.

The danger of interference with the freedom of the worker is much less in the case of the grant-in-aid in support of work freely undertaken by the individual. Even this, however, is not altogether free of danger. . . .

The National Research Council has, I think, fully recognized the necessity for support of university research without domination. The attempt has been made to make grants on the basis of the ability of the investigator, and to avoid the "project" outlook. On the whole, I think the attempt to avoid interference has worked out well. It is, however, essential for both the universities and the granting bodies to recognize the dangers and to be continually alert. A disturbing feature is that the more competent the granting body, the greater the *possibility* of overriding academic freedom. The universities are having enough trouble with professionalism and pressures for trivial courses and courses oriented towards industry, and it is important for granting bodies not to add to their woes. The monolithic structure of Soviet science is a good example of the dangers of over-planning science. In particular, such planning has greatly weakened the universities by transferring most research and graduate training from the universities to government institutes.

From the point of view of a granting body, it seems to me that the main principles should be:

(1) In so far as possible the basic responsibility should be left to the universities to provide people, space, and interest. (*But* what do you do if a university has some good people but is not interested in helping them?)

(2) The control of the grants should be in the hands of a body composed mainly of university people, to avoid bureaucratic action, or the suspicion of it. (*But* a group of university people, assembled as a committee, can be just as bureaucratic as any other body in trying to direct the programme.) I think, on the whole, that the Honorary Advisory Council of the National Research Council has been outstandingly successful in avoiding interference.

(3) It is important that the support be for people, not projects. In the first place, if a man is good enough to be supported, surely he is good enough to decide what to work on. The project approach will inevitably lead to a disastrous attempt to plan university research.

(4) The most important thing is to avoid planning, co-ordination, and the attempt by a committee to direct everything. But there are some difficulties. If it appears that Canadian science is deficient in a certain field, there are two alternatives: (a) Do nothing. This avoids interference with academic freedom, but ignores the responsibilities of the granting body. (b) Do something. This is interference with the principle of supporting some one to do what he himself wants to do. The only solution appears to be to try to stir up interest, and then give support. But here the pitfall to be avoided is that of supporting third-rate people just because you are interested in what they want to do. This brings us right back to the danger of the project approach. Also, it is certain that a gap in a particular field can be filled only by first-rate people entering the field. Lowering standards in a field of interest is certainly no solution.

(5) Finally, of course, it is important to cut red tape to a minimum. It is necessary to know if people with grants are doing anything worth while, but useless reports, rules, and so on should be avoided at all costs.

All in all, it is not easy to live up to these principles, but I feel that the National Research Council has done a rather good job of it, and I hope we continue to improve. The future will certainly be complex. More support is greatly needed, and I am sure it will be forthcoming. In recent years the rate of increase has been great, but so has the need, and I am afraid we are going to be faced with a continuing critical situation. . . .

Engineering and Technological Education

THE ASSOCIATION OF THE UNIVERSITIES OF THE BRITISH COMMONWEALTH
TORONTO, AUGUST 26, 1956

I GREATLY APPRECIATE the honour of being asked to make one of the opening statements at this discussion. I can only assume that I was chosen because I occupy a middle-of-the-road position as an engineer converted to pure science, as the head of a government organization barred by the British North America Act from any interest in education, but nevertheless a member of the National Conference of Canadian Universities, and finally, as a Canadian, accustomed to taking a mean position between the United States and the rest of the Commonwealth. I am aware that I can say nothing new to a group, whose members have been vitally concerned with the problem for many years, but I am presumably intended to provoke subsequent discussion by stating my views on as many controversial issues as possible.

In asking me to contribute to this discussion Principal James [Dr. F. Cyril James, Principal and Vice-Chancellor, McGill University, 1939–62] suggested that the subjects mainly of interest were three. First, the type of institution in which engineering may be taught, ranging from the technical college through the technological institute to the university. Secondly, the importance of so-called practical experience, again covering a wide range: from apprenticeships and the Higher National Certificate through "co-operative plans" to a training which is taken completely at the university. Finally, there is the question of specialization ranging from general engineering to specialization in the underlying discipline (mechanical, electrical, and so on) or specialization to fit a specific industry (refrigeration, nuclear engineering). There is no question that Principal James has put his finger on the important points. First, however, there are several questions which are basic to the whole discussion.

The Supply of Engineers

The subject of engineering and technological education is much in the news because there is a great drive to produce more engineers to overcome shortage which is reputed to be of grave proportions. The question of a shortage needs some examination before it can be accepted unreservedly. There has been, in North America, a great deal of talk about the seriousness of the situation, but the change from a boom to a slight recession has produced a surplus rather than a shortage of engineers. It should, I think, be realized that the ordinary laws of supply and demand are not operating freely. A large number of engineers are engaged on defence projects, directly or indirectly, and as long as each international crisis produces an expansion of development programmes without regard to resources of men and materials, an artificial shortage is bound to exist. One wonders also if some major projects would not advance faster if they employed fewer engineers. The shortage thus produced has led to hoarding, which is by no means expensive because the engineer does not cost much more than a good technician and can be used to do a technician's job. As long as a large fraction of engineers are being diverted into sales and administration one has grounds to doubt the seriousness of the shortage.

However, there is no doubt that real shortages exist in certain fields, and that there will be an increased real demand for more engineers in the future. The problem is therefore with us, but I do not think that, at least on this continent, the time is nearly as short as is made out. Above all, the education of engineers should not be regarded as a race with Russia, and schemes for expanding the supply of engineers should not be allowed to do so at the expense of the quality of their education. If we are not careful this cry for more engineers can do serious damage to universities, and can degrade the education of engineers to mere vocational training. Parenthetically, it may be remarked that the most serious situation in North America is the diversion of honours science students into engineering, with the result that the supply of engineers is being increased at the expense of the scientific foundation on which engineering rests.

The Purpose of a University

The question of the purpose of the university is a much discussed one. Should a university be useful? If so, how useful? Should it cater to what society needs? To what it wants? Is there something a little less tangible than the needs of society and the Gross National Product which should concern a university or other institution of higher education? In short, should not the university be seriously concerned with knowledge for its own sake? These are all questions which are far too big and far too controversial to consider *per se* here. Their implications, however, cannot be avoided in discussing engineering as a branch of higher education, and as this is the Association of the Universities of the British Commonwealth I presume it is only this aspect of engineering education with which we are concerned. . . .

In a technological society, of course, the university can hardly remain aloof, but it is certainly open to criticism when it allows itself to be made the instrument of mere professional qualification. From an engineering point of view the important point is whether the student leaves an institution of higher education with a basic education in the principles of his subject, or with a professional qualification to take his place in industry with the minimum of difficulty or expense to industry. This is the question which distinguishes education from vocational training. There is no doubt that in the past there has been a trend away from principles and towards the acquisition of a knowledge of technological trivia.

The main argument is the cry that the university should give society what it wants, or what it claims to need. This seems to me to be a totally fallacious argument because its acceptance would mean that the university will never lead society but will always follow. Such an argument also appears to rest on the premise that knowledge and scholarship have no intrinsic value and should not be pursued for their own sake. To reduce such an argument to absurdity it may be remarked that society may need, and certainly wants prostitutes but I doubt if this is a valid argument for a university degree in the subject. Actually such a degree would be far more in accord with humanistic traditions than many which are given in the *Commonwealth Universities Yearbook*, which takes 17 pages of fine print merely to give the titles of the degrees obtainable in the Commonwealth. There is no question that Flexner's comments are even more appropriate today than they were 30 years ago.

No one will deny that universities must make some concessions to the demands of society, but 17 pages of them seem to be too much. Also one has the uncomfortable feeling that the more outrageous demands are not always met with as much opposition as might be expected.

The Purpose of Engineering Education

This leads to the major question of the purpose of an engineering education. Originally, of course, engineering was purely empirical and rested on age-old experience in the "practical arts" rather than on a scientific foundation. When the Ecole Polytechnique was founded in Paris it was only a faint scientific beginning in the teaching of engineering. In such a situation it was not unreasonable to expect students to absorb a large amount of "practical" knowledge: principles could hardly be taught because in most cases they had not yet emerged. Today engineering is essentially merely a traditionally defined portion of applied physics, and what is needed by the student is more of the science underlying engineering, and far less empirical technological data. In short the "practical arts" have become "practical science." Unfortunately there has been the customary time-lag in reacting to these changes. Engineering today is hardly recognizable as the same subject it was 40 years ago, but the framework of engineering education has changed hardly at all.

It seems to me that today the basic principles of engineering education are identical with those of the rest of university education: the important thing is education, not vocational training. There is no reason why an engineering student should ever have seen a plant or a mine before he graduates. "Practical" knowledge can be acquired on the job and is certainly not a proper part of a university education.

From this point of view I do not believe that industry can or should participate in the students' education. So-called co-operative schemes for engineering education have little to recommend them and do positive harm in deflecting the students' interests from fundamentals. The practical aspect of the job should be acquired later and should not be mixed up with university education. I must say that I regard the Antioch scheme as one of the major errors in the history of university education. Neither a university student nor an engineering student is an apprentice craftsman no matter where he is trained, *if he is trained properly*. L. A. DuBridge, the President of the California Institute of Technology, has made some remarks about graduate research which are equally applicable to undergraduate education. "Research work is not a series of college courses; it is a way of life. And I do not believe that either a scientist or an engineer can become fully qualified for research and development work unless he has actually lived full-time in the atmosphere of a graduate institution, fully immersed in some phase of its research program. . . . Therefore I should like to urge universities to use their influence to stem the spread of so-called 'cooperative programs' in which it is assumed that graduate work and training for research can be achieved in only a few hours a week spent in a university classroom. . . ."

The Types of Engineer

The main problems facing the university are that the word "engineer" is being used to cover four quite different types of people, all of whom are in great demand, and that the desire for professional qualification is causing pressure from many groups. In fact, one of the more farcical attitudes of present-day society is the attempt by all conceivable groups to set themselves up as professions with the appropriate university degrees.

There are two groups who are not real engineers and who cause a great deal of trouble and loose thinking. The first are people such as those who obtain the Higher National Certificate. These are "practically" trained, and are top-flight technicians. They are vitally needed both in North America and elsewhere, but are produced only in Britain and other parts of the Commonwealth. Many more of them are needed, and their production may well be more important than that of more university-trained engineers. They have, however, nothing in common with a university-trained engineer, and their training is certainly not the responsibility of institutions devoted to higher education. It is a most inefficient and expensive procedure to use university graduates as technicians as is done widely in North America (apart from anything else they are usually very poor technicians). It is less expensive but equally inefficient to pretend as is often done in Britain that technicians are equivalent to university graduates in engineering.

The second group who present a problem are the engineering graduates in Canada and the United States who go immediately into administration, sales, technical services, and so forth. It is argued that an engineering education is a good general education for business life. Here there is ample room for argument, but I personally feel that an engineering training is not the best form of general education, and also is fantastically expensive compared with an Arts degree. It cannot be denied, however, that a major problem today is the total absence of any scientific or engineering knowledge in a so-called broad education. If it is assumed that, in spite of all objections, a degree in business administration is here to stay, it might be desirable to insert some general engineering into it, and produce a course which is a broader general education for technological administration.

There are two types of real engineer, but they need, in my opinion, very different kinds of training. The operating engineer, who will not be concerned with research or development, should nevertheless be educated broadly on scientific fundamentals. He will, however, because he is finishing his formal education at the first-degree level, need rather more technological information than the research man. Also, this should certainly be regarded as an appropriate place for the pass rather than the honours degree man. There will, of course, be a certain percentage of first-rate students whose interests fall into this category, but broadly the degree is aimed at the run-of-mine student. The student should not be regarded as qualified when he graduates, and no attempt should be made to give him in the university a substitute for practical experience.

Finally there are research and design engineers, and these are quite a different species. They need the maximum amount of engineering and scientific fundamentals and a high degree of specialization in them. These students need courses leading to an honours degree.

In all this the word "specialization" is, of course, being used in two completely different senses. Specialization is essential, and provided that it is based on the underlying discipline (mechanical or civil, for example) can be broadly based on scientific principles. Specialization aimed at a given industry is quite a different matter. In some cases such courses are merely a hodge-podge of odds and ends from various departments. My personal views are, I know, extreme but I can see little excuse for nuclear engineering or textile chemistry.

It seems possible that the recent swing in North America to "general engineering" courses on the one hand and engineering physics on the other is really separating out the various types of engineering education. General engineering would seem to be a satisfactory training for those who are not going to be professional engineers in a real sense. The ordinary engineering course is aimed essentially at the operating engineer, while engineering physics appears to be a satisfactory education for the research or design engineer.

The British and American Positions

It is an apparent paradox that in the United States efforts are being made in the California Institute of Technology, the Massachusetts Institute of Technology, and elsewhere to broaden the scientific base of engineering, and to expand technological institutes into full-fledged universities, while in England attempts are being made to produce more non-scientific engineers and to set up or split off separate technical colleges. In fact, however, it seems probable that the English and American objectives are the same and that they are trying to move onto common ground from extreme opposite positions. Thus in England there are essentially only people with the Higher National Certificate and highly scientific engineers, with no one in the middle. In the United States, on the other hand, there are essentially no technicians and few scientifically trained engineers, while virtually everyone is in the middle. The American effort to produce more scientific engineers and the English effort to produce fewer are thus really part of the same movement to obtain a distribution of people to cover the whole field. The different approach in the two countries is accentuated by the fact that prestige belongs to the scientist in England, but to the engineer in the United States. I must confess that I personally am not enamoured of the technological institute, and prefer to see engineering as part of a complete university.

Conclusion

It seems to me that, while there are many problems, the crux of the matter is to distinguish firmly between engineering education and vocational training: the role of institutions of higher education should be confined strictly to education.

The Life Sciences in Canadian Universities

AGRICULTURAL INSTITUTE OF CANADA
OTTAWA, JUNE 19, 1962

. . . I WOULD NOW LIKE to turn to what seem to me to be some of the peculiar features of the organization of the life sciences, particularly in universities. Part of the anomaly is due to the fact that agriculture is a very ancient craft, and that the old crafts have not taken easily the relatively new impact of science. Whatever the reason there are some distinct differences between the position of the life sciences and that of the physical sciences.

The general university organization of the physical sciences is that physics and chemistry are taught as such in single departments of their own in an arts and science faculty. The applications are clearly differentiated and lie in the faculty of engineering. Is it logical to have biology in the universities, and fundamental biology at that, taught and investigated in the faculties of arts and science, medicine, dentistry, pharmacy, forestry, agriculture, and so on? Perhaps we should at least be thankful that we do not have faculties of fisheries and wild life. It might be noted that in spite of its economic importance mining has remained a branch of engineering and has never risen to faculty status. It seems to me that the life sciences have suffered from their domination by medicine in the East and by agriculture in the West, and that the present situation is not ideal.

Logically, perhaps pure biology should be in the arts and science faculty and the applied aspects organized in medicine and agriculture; because medicine is basically mammalian science, agriculture might cover the rest, including plant science. It seems to me that it would be sound to avoid further subdivision at this level, and perhaps all the other fragments could be regrouped under these two umbrellas.

In general, at present, life science research in arts and science faculties is often weaker than that in agriculture and medicine. This is obviously because of popular financial support and stimulus to things of a practical nature. These factors produced a favourable climate for biology in the early days when university support for science was small, but their perpetuation is something which should carefully be considered. Certainly the present situation where most pure biological research is done in applied faculties is an illogical one.

Side by side with all this, and perhaps a result of it, has been a continual divisive effect in biology. Consider, for example, the relative positions of chemistry and of the life sciences. In the United States, to within about 10 per cent or so there are equal numbers of chemists and of life scientists. The actual figures for 1959 are engineers, 783,000, scientists, 313,000. Of the scientists there were 95,000 chemists, 28,000 physicists, 108,000 life scientists (30,000 medical, 41,000 agricultural, and 37,000 listed as biological). Yet chemistry has remained a single subject and is taught as such. It has been difficult to reorganize chemistry courses so as to modernize teaching, but, because only a single department in each university has been involved, it has been possible. Radiation chemistry, for example, is taught and investigated in many chemistry departments; in contrast it appears to require relatively drastic organization to undertake radiation biology.

This division of biology into fragments is proceeding at an accelerating pace: biophysics appears to be launched, biophysical chemistry is on the horizon. The position in physics is similar to that in chemistry: solid state physics has emerged as an essentially new subject and a very important one, yet no one in possession of their faculties would suggest an honours course in solid state physics. Somehow the essential unity of chemistry and physics has been preserved, but not that of biology. The situation is complicated by the fact that there seem to be at least four separate divisive influences at work: (a) by class or tradition—human, animal, or plant; (b) by scientific subdivision—physiology, pathology, etc.; (c) by profession—medicine, agriculture, forestry, etc.; and (d) by interdisciplinary stimulation—biochemistry, biophysics, etc.

One wonders if the interdisciplinary fragmentation is not fundamentally a reflection on teaching methods. In analogous cases there seems to me to be little doubt that geophysics emerged merely because geologists were too conservative to include sufficient physics in courses in geology. Similarly I think that the present popularity of courses in engineering physics is a reflection on the lack of modernization of engineering education. In biology could it not be argued, for example, that the rise of biophysics merely means that physiologists are not being taught enough physics?

As an example of the difficulties produced by such subdivision of biology, it is interesting to note that the National Science Foundation has only one grant panel for chemistry, but requires nine for biology: developmental, environmental, genetic, metabolic, molecular, regulatory, systematic, general, and psycho-biology. Another place where the disintegration of biology is causing real problems is with international organizations. The International Council of Scientific Unions has one chemical union at present under its wing and three biological unions. Ten years from now, if pressures are not resisted, there might well be 20 biological unions. The only other case in which there is a similar divisive tendency is in the earth sciences: again a science with an exploratory tradition. Here we have geology, geophysics, meteorology, oceanography, etc. The question is: Is the basic organization of biology wrong, and if so, is faulty organization causing harm? Though only an amateur, I suspect that the answer to both questions is yes.

Biology is a large field embracing much of physical science and in addition the organization, metabolism, reproduction, and so on, of living systems. But does the available knowledge of life science justify 10 to 20 times the subdivisions of chemistry and physics? I think not, and it is not good for biological science.

As far as the special position of agriculture is concerned, I would like to raise a few questions. Agriculture is certainly in a special position, for the agricultural "industry" comprises mainly individual and independent farmers. As a result governments perform almost all the "industrial" research. Where does this leave the university faculty of agriculture? The push for food and health has, I think, resulted in an undesirable influence of application and economic considerations on the relative amount of research in different fields of *pure* biology. There are, of course, similar effects in other fields of science but they are much less marked.

Is it an anachronism to have a faculty of agriculture? Is such a faculty an indigestible combination of different aims, including professional training, technician training, demonstration farms, provincial applied research laboratories, and a source of research and education in pure biology? If these functions are not really compatible, can anything be done about it? Are financial complications serious? They involve relations between provincial universities and provincial departments of agriculture. (Even in the East the Ontario

Agricultural College is affiliated with the University of Toronto, but financed by the Department of Agriculture.)

Apart from philosophical questions about the organization of biology and agriculture, it may be worth looking at the practical question of what can be done with the present organization. In research, I think there are three main questions. (a) Should more of the fundamental research basic to agriculture be done in universities rather than in the laboratories of governments at all levels? (b) Should such research be done in faculties of agriculture or in faculties of arts and science? (c) Should the more applied type of work be done in a university? In other words, should there be a movement of pure research from government laboratories into the universities, and of applied research in the opposite direction?

A further important question is whether there should be more support for research in agricultural faculties of universities. It seems to me that it would be wise to have federal support strongly increased through the interested department, especially if more pure research is required in fields bearing on agriculture. If such support could be given, and in the form of grants rather than contracts, I think much could be done to accelerate the rise in the research capabilities of agricultural departments of universities.

Finally, I would like to turn to the question of the position of the life sciences vis-à-vis the physical sciences. In this connection, it has sometimes been suggested that the life sciences have been neglected relative to the physical sciences, and in particular that the spending on, and support of the life sciences should equal that for the physical sciences. It seems to me that one thing can be said right away, and that is that there is no excuse whatever for equating the two. There is no obvious or logical reason why the "right" support is equality rather than ten times or one-tenth.

There are a large number of pitfalls. In considering national spending or technical people employed it is usual to consider research plus development and scientists plus engineers. Now it is virtually impossible to do development of the more expensive kind in agriculture, simply because by definition such development will constitute secondary industry. If one considers scientists, leaving out engineers, then about one-third of the scientists in the country are life scientists. If university support by the government in Canada is considered then nearly half goes to the life sciences. I do not think, therefore, that there is any *a priori* case of the neglect of the life sciences.

If the argument is put on the thoroughly subjective basis of importance we are on difficult ground. If we accept the humanists' outlook, life is all important. From a cosmological point of view it is, however, merely a minor phenomenon on a few of the less important astral bodies. I do not think it is worth following the philosophical question of importance further.

If the question is argued from the point of view of need it seems to me to be still thoroughly subjective. In the first place, industrially, agriculture is no longer in the dominant position it once occupied, and I think there are many arguments in favour of expanded research for secondary industry rather than primary. Also, though I am not saying that it is true now, it *is* possible to spend too much on medicine and to produce a situation where we sacrifice all the comforts of life merely to live longer. Who is to say, therefore, how much we ought to spend on medicine? I suspect that whatever decision is reached will be reached on emotional grounds. Further, any real decision on relative support or spending must be based largely on how good are the workers and the work. It would not be difficult to get a fair degree of agreement on who are the best 50 physicists in Canada. I defy anyone, however, to decide objectively whether the best 50 physicists are better or worse than the best 50 biologists.

I think, therefore, that it is very difficult to arrive at a decision as to the proper support for biology. I am sure that one should not be adversely affected by the many bad arguments which have been put forward for disproportionate support. I am sure that more support is needed, but I doubt if there has yet been a strong case to prove that the biologists are being discriminated against. I would, however, be willing to be convinced.

Finally, I would like to ask if it is possible that, apart from lack of support, real or imagined, biology *has* been held back by conservatism, by unwillingness to reorganize and to modernize teaching, by professionalism, and by the domination of the field by application? I am sure, of course, that the Agricultural Institute recognizes these problems, and this meeting is a sign that you are considering them. I am sure that there is much that you can do, and I wish you luck.

Where is Physics Going?

CANADIAN ASSOCIATION OF PHYSICISTS
HAMILTON, JUNE 17, 1958

IT SEEMS TO ME that one of the greatest dangers facing physics today is the growing social consciousness of physicists. This results in two things. The first is a developing lack of objectivity. It must be confessed that in many statements on both sides of the "fall-out" question, physicists have been acting as emotional advocates while pretending to be acting as judges of the physics involved. In the second place, such social consciousness and predilection for utility has meant an increasing ignoring of physics for its own sake. We all hope the physics will be useful—one might almost say more useful than it is at present! However, this usefulness will *not* be produced by physicists worrying about practical ends, and forgetting that physics is a branch of knowledge which deserves to be cultivated for its own sake.

An increasing illustration of such social consciousness is the preoccupation of physicists with world affairs. There is a strong group of nuclear physicists who seem to feel that all that is necessary to straighten out the world is to have it run by atomic scientists. From what I know of physicists, I doubt if things are quite that simple. It seems to me that the combination of this view with the view that physics can cure all ills is leading present-day physicists to lose their sense of humour and perspective and to take themselves far too seriously. I would hope that in ten years physicists might have regained their sense of perspective, and have realized that what they themselves have to worry about specially is the question of where physics is going, and the extent to which the freedom of physics, and of science in general, is endangered by many present tendencies. In particular, physics is in a difficult position, as the only science in which things other than those which are trivial are classified and kept secret.

The Big Machine

One of the major problems of physics which will have to be tackled seriously over the next decade or so is the problem of the big machine. Installations, especially for nuclear physics, have become fantastically expensive, both in initial cost and in upkeep and operation. This raises many problems, other than financial, and has some grave disadvantages. The most important has been the unfortunate rise of team-work as opposed to individual effort. One of the most accurate and frightening treatments of this subject is by William H. Whyte Jr. in *The Organization Man.* Apart from destroying initiative, the team appears to involve mainly the substitution in the supporting staff in physics research of psychiatrists for instrument makers. In the old days physicists stayed in the laboratory. Today most physicists over 28 have become executives. I don't even think that physicists today are as crazy as they used to be—or at least not quite!

Seriously, however, the Big Machine produces grave problems. Apart from its inhibiting effect on initiative, it invites outside interference because society can hardly be expected to give money in $20,000,000 chunks without at least asking what it is for. There is also the question of inflexibility. Once a group has wheedled a very expensive gadget out of society they have damn well got to keep right on using it even if their interests wander off to something else. In short, the scientist tends to become a technician, and to become a slave of the machine. There is also the question as to the effect on research students of being brought up in the shadow of a $50,000,000 gadget.

This raises the question whether the next 10 years may see some drift back to freer, more peaceful, more imaginative work in other fields. There is no question that fields develop as fads, especially in physics, because of their glamour. Is there too much nuclear physics today? As an outsider I would be inclined to say that there is far too much, and to suggest that in ten years, perhaps, solid-state physics will have replaced it as the major fad of the day. I am not at all sure that these violent swings in interest are a good thing, and I think that chemistry is lucky in being less subject to them.

The main question from my point of view is one that is not asked nearly often enough. Will physics still be fun in 1967 or 1977, or will it have become so organized into teams that it is no longer a challenging occupation for someone with intelligence and imagination?

The Location of Physics

This raises, I think, the central problem facing physics and the other sciences, but particularly physics, in the next 20 years. Can academic freedom survive if equipment becomes too expensive? Can you expect society to expend immense sums without asking for practical results? Can you avoid a "practical"—and ruinous—outlook on the part of physicists? Above all, can one preserve the university as a centre of pure physics, and if not, what happens to physics? I do not believe that science can retain much of its accustomed objectivity if nearly all pure physics passes into industrial hands. If you cannot believe this try reading the patent literature on a subject you know well. It is necessary to try to ensure that the university remain the centre of pure physics. But what will a large machine do to a university? Is it a nice thing to have a machine with 500 attendants on a campus, with whistles blowing for shift changes? Certainly experience along these lines so far has been rather dismal. On the other hand, is a machine 20 or 50 miles away a real part of the university at all? Can one educate a student there? I am not sure that you can, although of course you can *train* him, but experience along these lines has been equally dismal.

These are difficult questions, but somehow they must be solved. We are, I think, very foolish if we assume that we can keep all the good things in science and especially in university science, and at the same time progress by getting much more financial assistance. The last 30 years have brought a great increase in support for research—and a considerable loss of the freedom of science. Further support will have its inevitable hazards, and it is important to realize this and not merely to drift along until academic freedom has disappeared, which it may well have done by 1977.

The Canadian Association of Physicists in Ten Years

Finally, what of the C.A.P. ten years from now? Here I think one may make accurate predictions because its members are following fairly well along the path taken by the chemists. On this analogy, one can unhesitatingly predict that its membership will have increased greatly by 1967, and that the inevitable consequences of this will be: (a) meetings will be so much bigger than they are now that they will be hardly worth coming to, except for the purpose of meeting salesmen; (b) non-research people will inevitably be in control of the society, and there will be many complaints that the scientific sessions are not sufficiently suited to the practical man; (c) professionalism will have risen strongly and the society will be much more interested in the status of physicists and in surveys than it is in physics; and (d) there will probably be compulsory licensing, and no one will be allowed to do research in physics without a certificate from a dubious group in Toronto.

If you think this is foolish and unrealistic look at the chemists in Canada and look at societies like the American Chemical Society, with 85,000 members, and the American Physical Society which is catching up with the A.C.S. Then see what you can do, not to increase your membership, but rather to keep people out!

I apologize for making all these pessimistic remarks. I think, however, that there are many dangers coming, and that physicists *could* learn something from the chemists. Of course, they *won't*.

SCIENCE AND THE HUMANITIES

Science and the Humanities

UNIVERSITY OF WESTERN ONTARIO
FEBRUARY 15, 1956

. . . I HAVE TRIED to show something of the way in which science has developed, and to point out the things necessary for its continued development and the dangers to be faced. The development of science has, of course, had an effect on the universities and, in particular, on the position of the humanities in them. I have great sympathy for the humanities and their difficulties, although I am by no means convinced of the soundness of many of the arguments which are put forward. I would, therefore, like to make a few remarks on the question of science *versus* the humanities.

In the first place, it should be emphasized that as far as the universities are concerned the argument is a one-sided one. I am sure that today many professors in the humanities regret, perhaps with some justice, that science was ever let into the universities at all. On the other hand, no one in science will deny that from its point of view the university is the right place for it, and that the academic atmosphere, produced mainly by the humanities, is the only atmosphere in which pure science can flourish.

Are Science and the Humanities Incompatible?

I would like now to turn to the question whether science is the big bad wolf which is responsible for the decline in status of the humanities. Are science and the humanities so different in objectives and outlook that they are incompatible, and that one will drive the other out of existence? Is the scientist an uncouth fellow who stands for all that the humanists deplore? In fact, is he not only uncouth but proud of it?

It seems to me that on historical grounds there is emphatically no incompatibility between science and the humanities. In the first place science is merely a branch of philosophy, and is thus historically *one* of the humanities. Further, in the early days scientific investigation was carried on largely by humanists. The founding members of the Royal Society of London in 1660 included poets, writers, artists, and so on, all of whom felt that science was a proper pursuit for one of broad learning. The argument that the scientific method will solve all human problems, however, is no longer one which is taken seriously by any intelligent scientist.

It is sometimes suggested that the qualities necessary and the motives for the two pursuits are so different as to make for a complete clash of outlook. This is impossible to accept. There is no question that both imagination and scholarship are required for good work in either field. Further, because science consists of the pursuit of knowledge for its own sake, its motives would not seem to be too far from those of the humanist. One thing that confuses the issue badly is that the motives for the *support* of science are not necessarily those of the scientist.

The most frequent argument that science is on a low plane as far as human affairs are concerned is that the scientist is "narrower" in outlook. In this connection Alfred North Whitehead has remarked that "The antithesis between a technical and a liberal education is fallacious. There can be no adequate technical education which is not liberal and no liberal education which is not technical." I will return to this point later.

Finally, the argument is raised that the motives of the students differ, that of the science student being purely vocational. To this of course one can reply that Honours English is a good, and often used, training for writing advertising copy, and Honours Modern Languages for selling soap in South America.

Real Reasons for the Objections to Science of the Humanists

It seems to me that of the above arguments three are valid.

Experiment. The fact that experiments can be done in the natural sciences constitutes the major difference between the activities of the scientist and those of the humanist. There are thus in science a body of facts which have a real and independent existence. However, there is also the conceptual side of science, which still remains akin to, as well as being a part of, philosophy. Surely to reject everything in nature, except what goes on in the mind, as being of no interest or importance is a very *narrow* view, as well as being a hopelessly conceited one.

Applications. Here the argument is really with technology rather than science, and consequently is irrelevant. However, it may be remarked in passing that the idea that knowledge is objectionable because it might be useful is a peculiar one. Even the humanities try to argue that their pursuits bring happiness or interest to life.

When I listen to my friends deploring the decline of the humanities, I am surprised to find that the humanities are much less "pure" than is pure science. We have defined pure science as the search for knowledge for its own sake. If one is asked if Canadian chemistry is in good shape, the reply will be yes if there are competent chemists doing first-rate work in chemistry, and if there is a reasonable number of them. It would never occur to anyone to say that chemistry is in a decline because the average business man or housewife does not know chemistry well and does not prefer reading it to watching television. I would have thought that the same considerations would apply to the humanities. Their decline would presumably mean fewer scholars of high quality on the staffs of universities, and so on. I find, however, that what is really meant by the decline of the humanities is that the garbage collector prefers watching television to reading the classics. In short, the humanists appear to be interested not in scholarship, but rather in adult education. There is, of course, nothing wrong with the promotion of adult education, but if a subject is in a decline because the general public does not read if for enjoyment, then I submit that physics is in a much worse decline than are the classics. In fact, it appears almost certain that while the percentage of university graduates who read the classics may have declined somewhat (and I am by no means sure that it has), the percentage of the total population who read the classics (or anything else) has certainly risen greatly. If the humanities have declined it is a relative rather than an absolute decline. In a similar way the size of the staff of the Department of English in any university has certainly risen in the last 50 years, even if it has fallen relative to the Department of Physics.

Broader(?). The most important argument and the one most used is that the humanities are broader, that they deal with the "whole man," that scientists are narrow specialists, that scientists are often, to quote the *Massey Report*, "only glorified technicians, lacking any broad understanding of the field in which they labour."[1] It seems to me that this argument deals with specialization rather than with science, though it is usually camouflaged in such a way as to suggest that a narrow, specialized training in history will make a "broader" man

[1]*Report of the Royal Commission on National Development in the Arts, Letters and Sciences* (Ottawa: King's Printer, 1951), p. 138.

than a narrow, specialized training in physics. In any case there is a great deal of bunk, as well as some small amount of sense, in these arguments about specialization and its evils. Far too often the man who prides himself on his breadth is merely too lazy, or too stupid to be capable of really mastering anything. He falls back on the claim that he really does not want to know much about anything, but rather a little about everything. In other words there is an overwhelming danger that "broad" means shallow. Personally, I do not believe that anyone is really educated, no matter how broad he is, unless there are a few deep spots, and unless there is at least one subject in which he has reached the frontier of knowledge rather than the "short introduction" stage.

Apart, however, from specialization there is often the implication that there is more breadth to the humanities even if learning is confined to a narrow field. It is interesting to enquire into possible reasons for this view.

Is it because the humanities are useless? This does not seem to be a possible criterion. It excludes the social sciences, history, modern languages, English (because the inability to write English is stated as a reason for studying the humanities). It excludes the remaining subjects because they are stated to add to the enjoyment of living and are therefore useful.

Is it just that they are old? This also does not seem to make the grade. It would surely leave English in a precarious position relative to Latin and Greek. Also it would suggest that we should pay far less attention to such comparative upstarts as Greece and Rome, and devote our attention mainly to Ur, Egypt, and Assyria, or even to Neanderthal Man. It also raises the question whether science would be acceptable so long as it were out of date: teach nothing later than 1600 A.D., for example. Isn't this merely conservatism, and the conviction that what was a proper education 50 years ago is better than that today, just as 50 years ago 100 years ago looked better?

We have in fact been going to the dogs since the dawn of history. J. B. Conant, when retiring from the presidency of Harvard, remarked that much of what passes for appreciation of the arts and letters is really a combination of anti-quarianism and the old snob appeal of a "gentleman's education," and that those who appeal to such tastes do a positive disservice to the humanistic tradition which is in fact the tradition of the continuing triumphs of the *creative* human spirit. . . .

Is it motive, that is, education *versus* vocational training? It is often suggested that the character of the modern university is being destroyed by the "vocational" training of scientists replacing the old-fashioned broad education. This raises the question of why students took courses in the humanities in universities 200, or 400, or 800 years ago. I think there is no doubt that there were three main reasons. The first was that some few students had a real and overpowering interest in their subject. The same few students exist today in any field, humanities or science. Secondly, some students had ample private means, and no intention of ever working. A general education seemed the obvious thing. There are not many of these today. Thirdly, a large number of students were going to have to find a job, just as today. Today they tend to go into subjects where jobs are plentiful and careers attractive, and thus many head for science and engineering. In the old days the main jobs were the church, the law, and the government, and an Arts training was the vocational training of the day. It has been remarked that it is an historical error to believe that any institution of higher education ever flourished merely as a centre of disinterested studies.

It seems to me, therefore, that the argument for the superiority of the humanities turns out to be merely an argument against specialization. If brought out into the open as such it can certainly be judged more competently. From the vocational standpoint I think that there are two distinct and different questions which the scientist should ask the humanist. Does it do any harm to train in a university a scientist (who otherwise would have had no higher education)? In the first place, I don't think it does any harm to the scientist; I refuse to believe that scientific training is a brutalizing process which is intrinsically objectionable. I think that the motives of the scientist are sufficiently similar to those of other students that it can hardly do the humanists any harm to have him around. He may, of course, divert money away from the humanities, but on the whole I doubt if he has had this effect. The second question is much more significant. Does it do any harm to divert a student from the humanities to science? Here I think the position is quite different. Thus far there does not seem to have been much real diversion. The number of students in the humanities has not fallen; rather the increased number of students in universities has come about by an increase in the number of engineers and scientists without a corresponding increase in the humanities. There is the question, however, whether the next stage will be a real decline. One hears arguments which suggest that the students in the humanities represent a pool of wasted manpower. This is the real danger, and it is here that the real fight for the humanities should be made.

Materialism

One frequently hears it suggested that it is materialism which is responsible for cultural decay. Materialism, of course, has always been with us. It is only those with a moderately high standard of living who can ignore it. Certainly today more people have leisure and more people *can* read, and I think book sales would bear out the fact that more people do read.

It is, of course, true that material prosperity is far from being everything. It should not be overlooked, however, that material possessions enable one to be unhappy in a much more comfortable way. The happiness produced by reading a poem may be of a more spiritual type than that produced by gazing at a new refrigerator, but the latter is happiness just the same.

Conclusions

It seems to me that the attempt to put the humanities on a superior moral and intellectual plane is merely special pleading by those who are by no means disinterested. Frankly, I think no first-rate scientist minds being told by a first-rate humanist that he is a somewhat crude fellow. What is irritating and ridiculous is to see a third-rate scholar in the humanities, who has never had an idea in his life, trying to maintain a similar attitude of superiority.

The real situation seems to me to be that science *is* one of the humanities, although technology is not. Specialization is necessary and is intrinsically desirable, in spite of what the dilettantes say, but there is a real problem in trying to maintain breadth in spite of it. It is, however, just as narrow to concentrate on one branch of the humanities as on one branch of science.

The fact that on vocational grounds there has been a shift to science is merely the normal type of change which is always occurring. However, the real danger is that too great a diversion from the humanities may occur, leaving large fields of scholarship more or less abandoned. Certainly the decline of the humanities in this way must be prevented. It is undeniable that more support is necessary for those interested in the humanities, but it seems to me that mixing the question up with adult education merely confuses the issue.

Again, in a world in which science has a large impact upon human affairs it seems to me that broad education requires some appreciation of the workings of science. I do not feel that the ordinary method of cramming a few facts into the Arts student by means of one or two elementary or survey courses does any good at all. In fact I can see no reason why a well-educated person should not be almost totally ignorant of physics and chemistry. It seems to me that it is an appreciation of the methods, aims, and philosophy of science which is important, and that instruction at a more senior level on the history of science and technology might well be considered an essential part of a broad education.

All in all I do not feel that there should be any clash in outlook between the scientist and the humanist. One cannot blame the humanist for some bitterness over the relative magnitudes of the support in the two disciplines. It may, however, be some consolation if he realizes that the enthusiastic support of science by those who do not understand its objectives means increased interference with the freedom of science and raises many problems.

OPENING OF GEOLOGY-BIOLOGY BUILDING
UNIVERSITY OF WESTERN ONTARIO
OCTOBER 23, 1958

I have been present at, and have read the accounts of the openings of a number of arts and science buildings in recent years, and I have noted that there is something rather odd about the speeches which are given on such occasions. If when science buildings were opened the speaker pointed out that breadth of education required some exposure to the humanities and some attention to their welfare, and when arts buildings were opened it was correspondingly pointed out that a well-rounded education required some knowledge of science, and that it needed buildings also, then everything would appear to be in balance. In fact, however, the situation is quite different. There is no doubt that a well-rounded education requires some knowledge of both arts and science.

However, when science buildings are opened the speeches are invariably of the "yes, but" type, the speaker admitting the merits of science but cautioning his audience lest they condone narrowness and overlook the finer things of life. On the other hand, when arts buildings are opened the listeners are never cautioned against a narrow classicism, and told that science also has a place in education; rather, there is a general air of conscious superiority and self-esteem about the whole affair.

This one-sided attitude is rather reminiscent of that of the Prebendary of Durham who remarked about a century ago: "The advantages of a classical education are two-fold; it enables us to look with contempt upon those who have not shared its advantages, and it fits us for places of emolument, both in this world and in the next." . . .

UNIVERSITY OF ALBERTA
DECEMBER 8, 1961

. . . There is no question that society today needs more liberally educated people. However, a "liberal" education should be really liberal and include some knowledge of all those subjects which make a broad man, able to cope with the problems of the day. A liberal education must be redefined so as to include some understanding of science and technology or the facts of life. A liberal education has already been broadened to include economics and the social sciences, so why not the natural sciences? Such a broadening of the basis of a liberal education is not easy to achieve, but if it can be done it will mean greatly expanded opportunities for people thus educated. There is no question that many industries are today hiring engineers for administrative positions simply because the Arts graduate cannot understand the basic principles of technology.

If such a broadening of a liberal education comes about the position of the humanities in education in the future will be assured. If it does not come about there is a real danger of our following the Russian pattern and training more and more scientists at the expense of the arts and social sciences. Such a decline in the position of the humanities would be a tragedy, but if it occurs the humanists will have themselves to blame. . . .

SCIENCE AND THE NATIONAL ACADEMY

Science and the National Academy

PRESIDENTIAL ADDRESS, ROYAL SOCIETY OF CANADA
JUNE, 1955

THERE ARE CERTAIN DIFFICULTIES in deciding on a subject for the Presidential Address to the Royal Society of Canada. Past addresses have been of various types. In the first place, a general discussion of one's own field of work has often been given. This may be both appropriate and interesting in the case of an historian, an economist, or a medical man. The path which connects photochemistry with the general problems of society is, however, a rather tortuous one. It is true that the subject is the basis of both vision and agriculture, but I think it is better to stay away from it.

The address given by my predecessor last year is an admirable example of a second type. In it he reviewed the aims of the Society and its present status. I naturally hesitate to discuss a subject which was presented in such an able and scholarly way only a year ago.

A third procedure is to discuss general trends in a broad field. I have decided to combine the last two and to give what is essentially a supplement to Dr. Bruchési's address with special reference to science and its position as far as a national academy is concerned. In addition I should like to discuss in a general way certain trends in the position of science. Further, last year Dr. Bruchési explicitly compared the Society with the French Academy. As a scientist I should like to make my comparisons rather with the Royal Society of London.

It is, I think, worth pointing out that when the difficulties or shortcomings of our Society are under discussion two distinct aspects of the situation arise. The first is one over which we have complete control, namely, the things we do ourselves. Are our meetings properly organized? Are there things we are leaving undone? The second type of criticism, and the most usual one, concerns the actions of others. What is the status of the Society? Are we, as the senior learned society, consulted as frequently and given as much responsibility as we deserve? This latter aspect of the situation is one to which I wish to pay particular attention.

It is of interest to look at the position of science at the time of the foundation of the Royal Society of London in 1662. Superstition and an appeal to authority and tradition were just beginning to give way to the experimental method. As a result, most of the intelligent men of the day were attracted to the new method. Intelligence was required, but little formal knowledge, and hence real work could be done by the gifted amateur. As a result the members were a rather mixed bag. The active participants in discussions included Charles II, members of the aristocracy such as the Duke of Devonshire, "practical" people like Samuel Pepys, poets such as John Dryden, and those, amateurs or others, who gave a major proportion of their time to experiment. Two things are noteworthy. The first is that the Society's influence was great because it included most of the powerful men of the day: in short, science was fashionable. The second is that science was in an elementary and almost purely empirical stage. As a result there was no real distinction between the scientist and the inventor.

Over the three centuries which have followed the foundation of the Royal Society, three distinct types of change have occurred which have had a profound effect on the present position of science and of the national academy. These changes took place first in science itself; secondly, in the places where scientific work is done and in the people who do it; and, thirdly, in the relation of science and the scientist to society at large.

The first change of importance is that in science itself. At the time of the foundation of the Royal Society the only possible method of approach was a purely empirical one. Experiments were performed and the results noted. There was, therefore, nothing essentially different from the methods which had been used in the practical arts since the dawn of history. What was happening, however, was an attempt to bring order and objectivity into the empirical method. There was, therefore, little fundamental difference in outlook between the scientist and the inventor, or between what we would now call "pure" and "applied" science. However, the going was slow and not much of direct practical use resulted. Essentially science appeared to be a tremendously interesting intellectual pursuit, and highly suitable as an avocation for those of independent means.

As time progressed much was discovered, but the applications were not obvious in the early stages. As a result, by the beginning of the nineteenth century, science and invention had become quite distinct pursuits. To quote J. B. Conant:

> Popular credit throughout the nineteenth century went largely to the inventor, not to the scientist. This is particularly true in the United States but not much less so in Great Britain. . . . By and large the scientist in the nineteenth century was supposed to be concerned only with discovering nature's laws; the inventor was taking advantage of these discoveries for practical ends. The

attitude of James Clerk-Maxwell, the founder of the electromagnetic theory of light, towards the inventor Alexander Graham Bell was one of patronizing condescension (Clerk-Maxwell referred to Bell as "a speaker, who to gain his private ends, has become an electrician"). Professor Rowland of the Johns Hopkins University, addressing his fellow physicists in 1879, said "He who makes two blades of grass grow where one grew before is the benefactor of mankind; but he who obscurely worked to find the laws of such growth is the intellectual superior as well as the greater benefactor of the two." The scientist looked down upon the inventor and the inventor in turn was a bit contemptuous of the scientist; so too were some of the men of business who backed the inventors in successful enterprises.

In World War I, President Wilson appointed a consulting board to assist the Navy. Thomas Edison was the chairman; his appointment was widely acclaimed by the press—the best brains would now be available for the application of science to naval problems. The solitary physicist on the board owed his appointment to the fact that Edison in choosing his fellow board members had said to the President, "We might have one mathematical fellow in case we have to calculate something."

Another story illustrating the popular attitude towards science and invention in 1916 concerns chemists, not mathematicians or physicists. At the time of our entry into World War I, a representative of the American Chemical Society called on the Secretary of War, Newton Baker, and offered the services of the chemists in the conflict. He was thanked and asked to come back the next day. On so doing, he was told by the Secretary of War that while he appreciated the offer of the chemists, he found that it was unnecessary as he had looked into the matter and found the War Department already had *a* chemist.[1]

By 1920, however, the conceptual side of science had advanced to a stage where sufficiently detailed predictions could be made to render the application of science to industry well worth while. As a result, the stock of the scientist rose at the expense of that of the inventor. Today the inventor has mainly been replaced by the scientist in industry, and

[1] J. B. Conant, *Modern Science and Modern Man* (New York, 1953), p. 16.

the distinction between pure and applied science has largely vanished except for the important question of motive. The result has been a great alteration in the relation of science to society, and we shall come to this shortly.

The second important change over the years is in the people doing research, and in where they do it. As far as people are concerned we have gone a full cycle since the Middle Ages. In the early days (say prior to the seventeenth century) much of the research was done by professionals. By this I mean that most dukes and kings had a tame alchemist on the premises. He worked in the cellar and all that he did was kept highly secret. . . .

Once modern science commenced to develop in the seventeenth and eighteenth centuries research was mainly a pursuit of the amateur with private means. In the nineteenth century the scientist was usually a "semi-professional," that is, a university professor who was not obliged to do research, but whose career to some extent depended on it. Today research is largely a matter for the professional. The large laboratory with hired professional research workers is a recent development. It raises many problems because no one has yet really discovered how to administer such a laboratory without at least partially destroying the originality and initiative of the scientists.

A major change is also developing in the location of scientific research. Three hundred years ago when workers were amateur and equipment was simple most work was done in private laboratories. . . . I am sure today that many professors in the humanities regret, with perhaps some justice, that science was ever let into the universities at all. However, until recently almost all scientific work *has* been done in universities, and there is no question that from the point of view of science this is the place for it. It is only within the university atmosphere that a truly objective search for knowledge can be carried on.

Recently, however, with the increasing application of science, and the increasing cost of equipment, more and more research on pure as well as applied science is being carried on in industry. There is a great danger that this trend may destroy the dominant position of the university in science. There is also a danger that the increasing emphasis on technology may destroy the character of the universities themselves. Another factor which bears on the situation is the great increase in the magnitude of the scientific effort. All these questions raise many problems, and I shall return to some of them later.

A third major change over the past three hundred years is in the relation of science to society and, in particular, to government. I have pointed out above that alchemy and atomic energy both brought restrictions. The reason is, of course, obvious. The knowledge of how to manufacture gold or to produce atomic energy gives great power to those who possess it. The result is both an interest in science and a restriction on its freedom. . . .

In the days of the foundation of the Royal Society science did not seem to be of practical value. It was encouraged in a personal way by Charles II, but the government of the day did not have scientists in its employ, at least not as such. As a result science could happily be left to a society, and on the rare occasions when advice was needed the Royal Society was consulted. Again, because the Greenwich Observatory, and later the National Physical Laboratory, appeared to be useful but not absolutely essential there was no hesitancy in giving full or partial control to the Society. In short, governments were concerned with the practical side of things and were both willing and anxious to leave science to the control of an independent body.

Today, the importance of science is recognized and a steadily increasing number of scientists are employed by governments. As a result the national academies have to a considerable extent lost their position as authoritative government advisory bodies. This is, I think, inevitable because if science is important to government it is natural that the governments should set up their own scientific organizations and use their own consultants. In a sense, therefore, the position of the Royal Society of London is anachronistic. Even in it, however, there has been a gradual loss of authority, and the National Physical Laboratory, for example, is now in reality under the control of the Department of Scientific and Industrial Research.

It should, however, be appreciated that there is a great deal of difference between power and influence. There is no question that the power of the Royal Society of London has greatly diminished. On the other hand, it has to a large extent maintained its influence. This is due to several factors. It tries to select the best people and maintains its outlook as a working society, not merely as an academy to which it is an honour to belong. It has concentrated on doing a job, and not on insisting on its prestige. And, finally, it has shown an ability to move with the times by, for example, recognizing the need for having leading industrial scientists among its members.

The trend towards science in government appears to be inevitable so long as it is considered that science is of importance to the community. This trend is, I think, entirely reasonable. It is obvious that if large amounts of money are necessary for the support of science they will not be given without at least some curiosity on the part of government as to how the money is used. It seems to me that no national academy which maintains its independence can regard itself as a government adviser by divine right. The main point I should like to make is that there has been a drastic and inevitable change in the relation of the national academy to society and to the government as far as science is concerned.

Before discussing opportunities for useful work by a national academy, I should like to summarize a few trends which seem to me to be dangerous at the present moment. The first of these is the attempt to "plan" the effort in pure science. This stems largely from a confusion between the aims of science and those of technology. One must admit that this confusion is to some extent the fault of scientists who encourage it in attempts to get financial support for fundamental work. It is perhaps not unfair to say that some scholars in the field of the social sciences and even in the humanities have also encouraged such confusion under similar circumstances.

On this subject I would like to quote some remarks by Polanyi:

> The popular scientific books which I used to read as a child were mainly concerned with displaying the wonders of nature and the glorious achievements of science. They dwelt on the enormous distances between the stars and on the laws governing their motion; on the crowd of living creatures made visible in a drop of water under the microscope. Among the best sellers of the time was Darwin's *Origin of Species* and every new discovery throwing light on the process of evolution aroused intense general attention. Such were the topics and interests that came first to mind in connection with science even twenty years ago.
>
> It was not forgotten, of course, even at that time that science also provides a store of most useful knowledge; but this was not considered as its principal justification. New practical inventions like the electromotor or the wireless telegraph were thought to be merely occasional offshoots of advancing scientific knowledge.
>
> Today boys and girls who are interested in science are given a very different idea of it. They read books which profess that the primary function of science is to promote human welfare. . . . All these books emphatically oppose the view, generally accepted before, that science should be pursued for the sake of enlightenment regardless of its practical use. They have exercised a powerful popular influence which has been consolidated lately by the support of important organizations. . . . [The older] conception of science is still generally maintained by the academic profession; but it is no exaggeration to say that it is already beginning to be forgotten by the broader public, even though it was universally accepted by it only fifteen years before.[2]

[2]M. Polanyi, *The Planning of Science* (Oxford: Society for Freedom in Science, occasional pamphlet no. 4, 1946).

At first these remarks might appear to be incompatible with those of Conant, cited before, but this is not so. What is really implied is that 50 years ago it was important to overcome academic snobbishness towards the practical man. Today, on the other hand, the confusion of science with technology glorifies the scientist, but imperils the whole foundation of the scientific structure.

Another trend which arises from the same factors is the decline of scientific societies. At first sight this decline is by no means apparent. Societies are increasing in number, and their membership is expanding beyond all bounds, in one case to over 70,000. However, the increasing industrialization has brought about a great change in the type of membership. The majority of the members are apt to have no real interest in the advance of science. The societies tend to be professional bodies, and the meetings more and more to resemble the normal type of convention. The welfare of science itself, rather than that of scientists, seems therefore to be in the process of being handed back to bodies of the national academy type. There is no question that there will be an increasing opportunity for accomplishment in this field.

Another modern development has been the change in the type of support for scientific research in universities. Scientific research has become more expensive so that increased support is needed. This change has occurred, however, at a time of high taxation. As a result, endowment has become relatively less significant and universities have become more and more dependent on *current* support. This raises many problems. The best summing-up of the situation is the well-known remark that Stephen Leacock once made in an address at McGill. He declared that "the most glorious thing about James McGill is that he is dead," and went on to develop the argument that "the only good benefactor is a dead one." There is no question that endowment means, ultimately at least, that the funds are free from control, while the live benefactor, be he an individual, a corporation, or a government, is always just around the corner. The problem of academic support without interference has always been a serious one, and it will certainly become increasingly so.

Another problem of the present day, referred to above, is that of secrecy, and the attendant destruction of internationalism in science. This has been talked about so much that I do not want to enlarge upon it here. It is, however, worth mentioning once more its importance and the necessity of keeping alive the view that security regulations are intrinsically objectionable even if necessary in the disturbed conditions in which we live. . . .

In passing I should like to mention the large amount of loose thinking which is prevalent on the moral obligation of scientists to prevent the misuse of science. Science merely furnishes the knowledge. This can and will be put to whatever use the society of the day dictates. The moral obligation of the scientist is thus no more, and no less, than that of any other member of society. Further, one cannot develop science piecemeal, and a return to seventeenth-century weapons means a return to seventeenth-century public health.

To sum up, recent developments in the relation of science to society have certainly diminished the real or potential power of the national academy as far as science is concerned. It seems to me, however, that if such a society can become a real working group, rather than a congenial club of scholars with distinguished pasts, there is still much to be done. The upholding of the freedom of science, of the university, and of the scientist seems to be one of the main problems with which the society might concern itself. Much can, I think, be done, provided that vague, high-sounding objectives are ignored, and attention is paid to real problems open to attack.

Finally, I think it is important to consider why the things which worried Dr. Bruchési in his address last year are so different from those which I have discussed this year. Our society includes the humanities, the social sciences, and the natural sciences. In the case of the natural sciences, there is no lack of popular recognition of their importance, and there is considerable financial support. The main danger is not lack of recognition but rather that recognition based on misunderstanding may destroy the whole structure of scientific freedom. In the case of the humanities, however, the main difficulty is the lack of recognition and support. The social sciences appear to be in an intermediate position, and will soon begin to find that increased support brings with it increasing interference. The possibilities for useful activity by the Society are thus quite different in the three cases. I think that there is much that is worth doing by the Society but it is essential that the different groups should realize the different nature of the problems in their respective fields.

SCIENCE AND SOCIETY

The Impact of Society on Science

PURVIS MEMORIAL LECTURE
SOCIETY OF CHEMICAL INDUSTRY
MONTREAL, NOVEMBER 27, 1957

I HAVE CHOSEN an "inverted" title for this lecture, because it seems to me that, if the impact of science on society has been spectacular and both beneficial and horrible, the impact of popular ignorance on science has been equally spectacular and similarly has its horrible aspects. The growth of science and of the importance of science are major features of our day, but we are faced with many serious problems because of popular ignorance of what science is or does: secrecy, manpower, sputnik, the distinction between science and technology, and the relation of science to the humanities are examples which all give scope for maximum misunderstanding.

The dangers of the lack of understanding of the aims and methods of science have, of course, been widely recognized, both by scientists themselves and by a vocal group who make their living by popularizing science. The difficulty is that most of these efforts, especially by the second group, are irrelevant because they adopt the "ain't science wonderful" approach.

A lecture in a church basement on "new wonder-drugs," "the marvels of plastics," or "chemistry in the service of man" may be appreciated by the audience and may give them a few disjointed facts. It makes, however, no contribution to the main problem of developing an increasing understanding of the methods, aims, and scope of science. In fact, it seems to me that many popular lectures on science lead to a complete confusion of ideas, and produce in the minds of the public the idea that science operates by producing an interminable and random succession of gadgets. This also serves to distract attention from the fact that it *is* essential that the leaders of our society have some appreciation of the implications and methods of science. The "wonders of science" approach represents an over-selling of science and its usefulness which in the end will be recognized as such by the public, and does more harm than good.

It seems to me that the most outstanding example of ill-informed public opinion is that which has surrounded the launching of sputnik. It is this which has tempted me to choose the present subject, and I would like to discuss some of the factors which contribute to misunderstanding. . . .

In connection with the rise of technology in Europe, Blackett, in his recent Presidential Address to the British Association, has given some figures which have considerable significance. It would appear that in the East, where the rise of technology has not occurred, the standard of living, that is, real income per capita, was about the same as that in Europe three or four hundred years ago, and has not changed since. In the West, however, industrialization has resulted in a rise by a factor of about 10 in Europe and 15 in America. The reasons why Europe suddenly "took-off" after being dormant for so long are somewhat obscure. There is no doubt, however, that the *continued* rise in living standards has been due entirely to the application of science to technology. If it were not for such application of science our living standards today in the West would probably be about the same as present levels in southeast Asia.

Those who criticize the materialism produced by science and pine for the good old days when education was classical and uncorrupted by science are inclined to overlook these facts. After all, the classical tendencies in education did not help the majority of the population much in those days, because they could neither read, write, nor get enough to eat. It is a curious fact that when we consider olden times we always unconsciously associate ourselves with the top 1 per cent income bracket.

As far as the impact of modern technology on society is concerned, there is a great deal of inept discussion. The question is frequently asked, "How does technology affect society?" in somewhat the same way that one might ask, "How does measles affect society?": in other words, as though technology was a quite extraneous influence. Now, in fact, society and technology involve the same people and the same things, in the sense that technology is merely the sum total of what everyone, or almost everyone, does for a living. It should also be emphasized that this has always been the case. The impact of technology on society is therefore merely the impact of what society does upon itself. It is by no means an outside, unpleasant force exerted on society by a few engineers and scientists, but is the collective influence of everyone's actions.

The real problem is not technology itself but rather technological innovation: this is what upsets the peaceful course of our lives. Technological innovation has, of course, always been with us. The problem in recent years mainly has been not the increase of technology, but the rapid rise in the rate of technological innovation, and it is this rise that has made our living standards what they are today. It often seems to be suggested that such technological innovation is a juggernaut which crushes society in its course, and that society has no power to combat or modify its effects. This is, in fact, the exact opposite of the true situation.

Science has developed an increasing understanding of nature. As this understanding develops there is an increase in the pool of natural knowledge on which technology is based. The technological innovation that results, that is, what is invented, is then a matter for society to decide. Far from technology forcing itself on society, it is society which ultimately controls technological innovation. A given technological innovation is therefore by no means inevitable, but is a definite and deliberate choice of society; whether society exercises this choice in a sensible way is, of course, quite another matter.

The main point is that the argument is often made that technological advance has a great influence on society, with the tacit implication that society has little influence on the direction of technological advance. Actually the direction of technological advance is apt to be due far more to advertising and sales and promotional efforts than to the efforts of scientists and engineers. For example, it is equally possible from a technical point of view to have automobiles get longer, more ornate, higher powered, and more expensive, or to have them get more durable, cheaper, and more convenient. The direction of the development is decided by the public under the influence of mass media of communication. Science has the major influence on what is possible, but only a minor influence on what is, in fact, done.

The crux of the matter is that the development of scientific knowledge, and the potential technological advances which may arise from it, have given society the chance, for the first time, to make decisions on many matters which in the past have been largely or totally beyond its control. For example, in the past the population of the earth, or of any given part of it, has been largely dependent on disease, fertility, and so on. Today for the first time we have the information, the ability to control disease, and so forth, to enable us to make effective decisions about population. There thus arises the question whether society is willing to make any decisions at all about the matter, as well as how intelligent such decisions might be. It is, however, essential to realize that potential technological innovation is offering society freedom and not the reverse. At the same time it is making it essential for society to seize the opportunity to make decisions, and the future will bring up many awkward questions.

It is perhaps worth emphasizing also that there is a great deal of loose thinking on the question of the moral responsibility of science and scientists for things like nuclear weapons. All science can do is to increase the fund of natural knowledge and thus increase our potential control over our environment. What society does with this power is a social problem. There is no advance in the arts which cannot be used for objectionable as well as desirable purposes. If writing had never developed there would be no yellow journalism and no comic books, but I doubt if my humanist friends would agree that the development of writing was unfortunate. We have always been in a dangerous situation, and one beyond our control. What frightens us today, however, is not that the situation is worse, but rather that for the first time we have the elements of control within our grasp, and do not choose to use them.

The interlocking of science and technology has led to considerable confusion about the aim of science. It is, of course, the function of science to enquire into the workings of nature, and the application of such knowledge has become the mainstay of technology. The natures and motives of science and technology are thus distinctly different although in many cases their methods may be very similar. Science is thus in a dual position as part of a humanistic education (after all, it *is* a branch of philosophy), and as the basis of technological development. One danger of the importance of science to technology is that science in its own right as a branch of knowledge is apt to be overlooked or minimized. It is, in fact, the ignoring of science in its own right which has been responsible for drumming up a purely bogus clash in outlook between the scientist and the humanist. An example of the lengths to which such misunderstanding can go is given by a prominent local newspaper whose Washington correspondent recently defined *pure science* as "the basic knowledge underlying weapon development." The importance of science to developing weapons cannot be over-emphasized, but it is equally important to realize that science is not merely a military adjunct.

There is a curious ambivalence in the attitude of society to the development of science today. On the one hand, there is a passionate devotion to the results of the applications of science, leading to demands for more scientists and more science. Along with this, however, is a vocal but rather vague feeling that scientists are narrow, uncouth, and ungentlemanly, and that their thoughts about nature are in every way inferior to the corresponding thoughts of the Greeks some two thousand years ago. Scientists are also suspect because they have indulged in the crowning vice of specialization. I would like to discuss these two questions briefly. . . .

As far as specialization is concerned, it is obvious that only in a preliterate society is there no specialization. In fact, a professor of classics is a highly specialized product. Actually most arguments about specialization are dubious: they imply that no one can be broad unless there is no subject about which he really knows anything. My own feeling is that no one can be broad without a few deep spots. It should also be realized that in one sense the humanist *can* be narrower than any scientist. No one can get a degree in science who is unable to read or write. He must have a slight acquaintance with history, some modern language, and so on. It is, however, *possible* to obtain a degree in the classics without any exposure to science at all.

The argument of science *versus* the humanities is not a profitable one to follow further. There are no real clashes between the two, but only some rather vigorous special pleading on both sides. There are, however, two points which need emphasis. I do not believe that a B.A. necessarily produces a "broad" man or a B.Sc. a "narrow" one. Personal characteristics surely count more than formal training, and I refuse to accept the idea that all scientists should be classed as second-rate citizens. . . .

The change in man's civilization, outlook, and knowledge in the last 300 years constitutes a revolution as great as that of the Golden Age of Greece. Can one ignore all of this and still have sufficient breath of education to decide where society is heading? The major new factor today is man's ability to exercise control over his environment. It is difficult to see how a man can express contempt for his environment and all knowledge of it and still claim to be educated. In short, can you deal with the "whole man" while neglecting his environment altogether?

These questions pose a serious problem for future education. Somehow it is necessary to give future leaders of society some general idea of the aims and methods of science. The frequent lack of any such idea is a dangerous factor in our present situation. The problem, however, is not easy, and certainly is not to be solved by cramming Physics I and Chemistry I down the throats of Arts students. It is, however, a problem which must be solved somehow. It seems to me that the major step is to convince Arts students that some knowledge of the philosophy and methods of science is necessary to round out their education. The main thing is to overcome the attitude which has led to the definition of a scientist as a man who knows a little about the humanities and is ashamed that he does not know more, while a humanist is one who knows nothing of science and is proud of the fact.

The confusion over science reaches its maximum over the

questions of scientific manpower, and of the status of Russian science. In the first place, the whole concept of manpower as a commodity to be bought, sold, produced, and consumed is objectionable when applied to the end-product of a university education. It is also a dangerous concept in that it implies the unimportance of quality, and the importance of mere numbers. It is also obvious that the question of the education of engineers and scientists should not be treated as a race with Russia. The real questions are whether we are educating enough engineers and scientists for our needs, and whether we are educating them well. If we are worried about the pace of Russian science the answer is not to copy Russia, but to make sure that we are giving the maximum encouragement to Canadian science.

At the present time the shortage of engineers and scientists is not acute, except for the shortage of first-class men which always exists and always will exist in every profession. The real difficulty is that an acute shortage would develop over the next few years as the economy expands if the present rate of production were to remain constant. However, over the same period, because of birth-rates and other factors, there will be a tremendous increase in university enrolment.

Provided that the universities can expand to take care of the increased number of students, the increased supply of scientists and engineers will automatically be forthcoming. The major problem, however, will be, in the face of very large numbers of undergraduates, to maintain the quality of university teaching and in particular to keep up both the quantity and quality of university research.

The conclusions to be drawn are twofold. The first, and negative one, is that we do not need to use propaganda to coax students into science or engineering, or to coax them away from the humanities. The second, and more serious point, however, is that this argument is predicated on the assumption that students who wish to take engineering and science can be taken care of by the universities. This is the critical point. As a people we give very much less support to our universities than do other countries with a high standard of living. The critical factor in the whole manpower situation is simply whether we choose to give the universities the support they need or not. There is nothing to the situation but this, and the spate of conferences on the subject have, if anything, confused the issue rather than clarified it. . . .

The recent launching of earth satellites by Russia has indicated the great confusion about the problems involved. In the first place, there appears to be surprise that Russia has been able to accomplish anything, scientific or technological, in advance of the West. There are, of course, no grounds for surprise, as everyone connected with Western science has been pointing out for the past ten or fifteen years. Because the Russians have put a major effort into science they were bound some day to reach a reasonable degree of equality with the West. After all, the attempt to expand and improve Russian science has been going on for 40 years. Everyone admits that the United States is a major scientific "power" today, but what was American science like before World War I? If America can make such strides in 40 years, why not Russia?

This attitude of the inevitable superiority of the West has had unfortunate consequences. The present tendency to treat scientific advance as a race, with emphasis on nationalism, is most objectionable and is contrary to all that has been good in scientific internationalism for the past three or four hundred years. Also, the attitude that Russia cannot produce anything original has done the West an enormous amount of harm. It has led people to regard the keeping of secrets as of major importance, and has resulted in grossly exaggerated security measures. In many cases this has meant concealing from our friends things which were already obviously known to our enemies. One good result of the present furore is the sign of a relaxation in secrecy.

Though we obviously have cause to worry, it is important that we worry about the right thing. The pendulum has swung so far that within a few weeks of a time when Western scientific superiority was taken for granted, we hear statements that Western science is lagging far behind. This is, of course, complete rot. There is more to science than merely developing weapons. There is no question that on the whole Western science is still ahead of Russia. What has been shown by recent events is twofold: first, that Russia can compete on equal terms in science, and second, that by pushing a technological development she can get ahead of us if we don't watch our step.

What we need to do is to consider whether we really are interested in doing everything we can to develop Canadian science. If we are, we will have to see that we no longer lag in the provision of support for science in universities and elsewhere. The development of Canadian science has been striking in the last 20 years, in spite of relatively poor facilities. With reasonable support there is no reason why the future should look dark, but we will have to pay for it.

Implications of the Atomic Age

COUCHICHING CONFERENCE
AUGUST 13, 1955

I WAS ASKED by the programme committee to discuss the implications of the Atomic Age, including "consideration of both destructive and constructive possibilities of nuclear energy. In short, the meaning of an atomic age for a world whose peoples do not yet comprehend its full significance." It seems to me that my major task is to try to place nuclear energy in its proper perspective in relation to the general development of science and of technology. So many wild statements have been made about the "Atomic Age" that the public is, through no fault of its own, in a state of complete confusion.

It is worth going back for a moment to consider the development of science and technology over the last few hundred years. . . . With the development of the so-called scientific method, which is essentially merely an attempt to be objective, curiosity about natural phenomena became widespread. This led to an attempt to explain many things which were well established in the traditions of the industrial arts. In other words, science began to have some applications to technology. At the same time the improvements in technology raised points of scientific interest, and led to improved apparatus and techniques. Science and technology thus reacted upon each other, and development was rapid. This accelerating development of science and technology is frequently said to have culminated in the industrial revolution. The word "culminated" is unfortunate, however, because it implies a sharp, definite stage in development. In fact, there was no such well-defined stage, but merely a continuous and accelerating pattern of technological development based on scientific advances.

If one looks back at the developments of that time, it is interesting to note that the steam-engine came along at just the right moment to supply power for the newly developed machines. This was by no means accidental. In general, when there is sufficient need for a device it will be invented, and the industrial revolution was the result of the simultaneous development of the machines and the power necessary to drive them.

It is worth noting that, looked at from the perspective of today, we realize that it was the application of machines to industry in general that was important, not merely the specific invention of, say, the cotton gin. At the time, however, it was the cotton gin which was blamed for much of the social upheaval. In other words, there was a definite tendency to confuse the symptom with the disease. Today, however, we can recognize the fundamental part of the process, and no one refers to the period since the invention of the cotton gin as the "Textile Age."

It should also be emphasized that the process of mechanization of industry was a steady, continuous one. It was merely the rapidly accelerating pace which confronted society suddenly with the problems of the industrial revolution. The social problems may have appeared suddenly, but the technological developments which led to them were part of a continuous process.

The development of technology based on science has continued to accelerate. One cause has been the development of industrial and military research. It is only in the last 30 years or so that the industrial research laboratory, or the professional research worker, has come into existence on any real scale. This has led to a great increase in the use of automatic devices, and is causing a further great change in technology. The important point, and the one I wish to emphasize, is that the production of nuclear energy is merely one result of this development. The essential developments are not the discovery of atomic energy or atomic bombs, but the mechanization of warfare and the "automation" of industry. There is really no more excuse for talking about the "atomic age" than there would be for calling the industrial revolution the "textile age." I think that much of our thinking has been distorted by regarding atomic energy as an isolated and unexpected development, rather than as merely a result of the continuous development of science and technology. That it is a striking development no one can deny, but it is the symptom of the times, not the disease.

On this basis we now come to a more detailed consideration of the military and the non-military applications of nuclear energy.

The Military Aspects of Nuclear Energy

There is, I think, considerable confusion about the moral aspects of developing weapons, and the relation of science to technology, military or civilian. It is the function of science to investigate natural phenomena, with no reference whatever to the question of utility. In other words, science is concerned with the question, "Why?" Technology, on the other hand, applies existing scientific knowledge to accomplish specific ends, and is concerned with the question, "How?"

It is not possible to develop science piecemeal, or with a specific end in view. As a result, at any given time society has at its disposal a certain body of natural knowledge. It can use this for any desired purpose, military or civilian. Weapons, therefore, are a part of technology, and as technology develops so will weapons. One cannot, for example, have seventeenth-century weapons without having seventeenth-century public health. As an example, it may be remarked that virus research benefited greatly from the development of the electron microscope. It is no exaggeration, therefore, to say that the recent development of a polio vaccine stems from the same basic experiments which led to the atomic bomb. The idea of calling a halt to the scientific development of weapons is therefore an impossible one. One could do this only by calling a halt to all technological progress, including medical.

The terrifying thing that society is up against today is not, therefore, basically the atomic bomb, but the mechanization of warfare. The modern development of science and technology has given us a control over nature which we have never had before. This control can be used equally well for destructive as for constructive purposes. Our powers of destruction are thus greater than ever before. Further, because there is no reason to expect a halt in technological development, it is obvious that our destructive powers will be greater tomorrow than they are today. The atomic bomb is thus not an isolated, horrible development. It is merely the outstanding example at the moment of a continuously developing technology used for purposes of destruction.

It is interesting to speculate on what would have happened if nuclear energy had not proved to be possible. My personal feeling is that it would have made very little difference. With modern development of electronic techniques it seems certain that guided missiles with conventional explosives would ultimately have had destructive powers not far short of what can now be accomplished by the atomic bomb. The development of the atomic bomb probably brought destructive power to a stage today which otherwise might have taken 20 years to reach; it did not, however, introduce anything really new into the situation. Our dilemma today is that modern technology has placed vast destructive power in our hands. The atomic bomb is merely the present outstanding example of this.

It is noteworthy that even medical advances, which we certainly think of as beneficial, have strengthened the hand of a destroyer. In the past epidemics tended to keep victorious armies in check. Today medicine enables destructive forces to operate continually.

Peaceful Aspects of Atomic Energy

There are three aspects of the peaceful use of atomic energy which need discussion. The first are the ideas which come from the lunatic fringe of humanity. An example, quoted from the press of a few weeks ago is: "A supercharged vacuum cleaner manufacturer predicted Wednesday that a self-operating nuclear-powered vacuum cleaner with a built-in magnetic memory to guide it around a room may be a standard household appliance in 10 years." Such claims are, of course, merely amusing. Unfortunately, serious suggestions which are almost as ludicrous frequently appear from more reputable sources. Thus, a few weeks ago the editorial page of one of the leading Canadian newspapers carried the following: "The advent of atomic energy has made it possible, for the first time, for all the world to enjoy abundance. Even deserts can be made fertile, and polar regions be made to give up their mineral treasures." It must be admitted that atomic energy has certainly made the imagination fertile. The unfortunate thing is that such statements are so widespread that the public has been given a false impression of what is to be expected from the development of nuclear energy.

The second aspect of the peaceful use of atomic energy is a real one, but it has been very much overplayed. This is the use of radioactive isotopes as labels by which processes may be followed in research. The medical applications of isotopes have been much exaggerated, and the public is told that enormous progress, otherwise impossible, will be made in medical research. The use of isotopes in research, medical or other, furnishes a new technique. Any new technique offers opportunities for new types of investigation. Isotopic experiments will undoubtedly contribute much useful knowledge, but so does any other technique, and the progress of research is largely the result of such new techniques. There is no doubt of the utility of such isotopic investigations, but there is no question that their importance, great as it is, has been vastly over-emphasized. The mere application of isotopic methods is certainly not going to solve all our problems in science, medicine, and technology. (I may remark here that I have personally made much use of such methods, and am certainly not prejudiced against them.)

Other sidelines of nuclear energy include such things as the cobalt bomb for cancer treatment. Again, this is a most important tool, but by no means revolutionizes medicine.

Nuclear Power

The really useful application of nuclear energy is to furnish power. This is the only use which justifies excitement over the possible effect of nuclear processes on society as a whole. It has a very important bearing on the future of society, but it is far less glamorous than the imaginative remarks we have been referring to above. . . .

There seems to be no question that the major technological event of the next 50 years will be the increasing and widespread use of nuclear power. There is also no question of the very great importance of pushing this development forward with the greatest effort and speed.

Social Consequences of Nuclear Power

It appears, therefore, that the consequences of the development of nuclear power are of the greatest importance. They are, however, essentially negative consequences, involving the prevention of a serious power shortage rather than the production of new devices or material. Nuclear power may be expected to have an important secondary effect as far as standards of living are concerned, by preventing a shortage of power which would otherwise slow down the development of technology. It is, however, doubtful if the social consequences will be anything like as great as those which followed the development of the internal combustion engine. That enabled the development of the automobile and the aeroplane and produced a real revolution in means of communication. Any similar development which comes from nuclear energy will be only secondary, permitting other technological developments.

There is, however, one other feature of nuclear power which may have far-reaching effects. This is the fact that essentially mobile power will result. Nuclear power stations can be set up in areas poor in power, or remote from civilization. . . .

Conclusion

To sum up, it seems to me that there is no question of the difficulties caused by the advent of the atomic bomb, or of the importance of the development of nuclear energy. These are, however, not the basic problems, but merely aspects of a much broader one. The real problem facing society is to adjust to the increasing advances and demands of technology, whether in developing weapons or in the civilian aspects of the situation. The problem is essentially merely a continuing phase of the industrial revolution, and of man's increasing ability to control and guide the forces of nature. It seems to me that there is not, and never will be, a period which can justifiably be called the "Atomic Age."

GOVERNMENT SCIENCE

Government Science and the National Research Council

MONTREAL CANADIAN CLUB
DECEMBER 8, 1958

THERE ARE TWO MAIN FACTORS which have affected the development of industrial research in Canada. In the first place, in a pioneer country primary industries obviously develop long before any appreciable amount of secondary industry exists. As a result, in Canada first-rate facilities for research in agriculture and in mining developed long before industrial research as such got going at all. This is the normal course of the development of research in a country as it becomes industrialized. The second factor is that, because of the proximity of Canada to the United States, and because of the financial relationship of Canadian firms to those in the United States and Britain, research normally is done by the parent organization outside the country. (There are, fortunately, a few conspicuous exceptions.) The over-all result, however, has been that Canadian industry has been dependent largely on research done in Britain and the United States. This has given Canadian industry the enormous advantage of fully developed research organizations and highly advanced technical know-how, but has nevertheless resulted in scientific colonialism at a time when political colonialism was disappearing. It should be emphasized again that this is merely the normal course of development in a relatively backward country next door to the most highly industrialized nation on earth.

The result is that by comparison with the United States or Britain relatively little industrial research has been done in Canada by industrial organizations. On the other hand, the situation has long been appreciated by governments, federal and provincial, with the result that a great deal has been done by government agencies for industry. It is particularly noteworthy that in those few cases where Canadian firms are large by American or British standards they often, but by no means always, do a considerable amount of research themselves. The pulp and paper industry is an example. It is also noteworthy that as far as primary industries are concerned Canada's position as a producer of raw materials has resulted in government research programmes in mining and agriculture comparable to those of any country in the world.

It is frequently emphasized that we are a small country and hence cannot be expected to do much in the way of industrial innovation and research. However, it should be remarked that we have a much larger population than Denmark and Switzerland and many other countries in which appreciable research *is* done. We are faced with real problems because of our geographical size, but we are by no means helped by our attitude. It is a Canadian custom to boast about our natural resources for which we can claim no credit and should rather acknowledge our good fortune. We rarely take the attitude that we should be able to do some things better than anyone else. (Hockey is almost the sole exception.) In many ways the magnitude of our resources has diminished our initiative and allowed us to coast along on our good fortune. This attitude is, of course, a natural historical inheritance, but it is time that something was done about it.

At the same time it should be emphasized that the situation is far from easy, and that it must be approached objectively, not emotionally. Strung out in a long, thin line next to a much larger and more highly industrialized neighbour, afflicted with relatively small markets and the competition of industrial giants, the situation which has developed is a logical one. We cannot change it overnight, but rather must move slowly but steadily in the direction of more self-sufficiency in research. We cannot escape being scientifically (and industrially) dominated by our neighbour, but we must try to do something for ourselves and must make sure that what we do is first-rate by anyone's standards. The most encouraging feature of the situation has been the strong trend in the last ten years towards some degree of Canadian self-sufficiency in research even when the company is controlled from abroad. This has been particularly marked in the chemical industry, but conspicuously absent in certain other industries which shall be nameless. It is a natural tendency because the size of Canadian companies has been growing steadily. It has been very encouraging to see more and more Canadian companies starting research laboratories for the first time and the substantial growth of some of the older laboratories.

It is worth pointing out that there has been a complete revolution in the relation of science to society and to government in the last 50 years or so. Prior to that time science was a game performed by amateurs for their own intellectual satisfaction. Invention, on the other hand, was performed by practical men steeped in the traditions of their crafts, and the crafts advanced empirically and exceedingly slowly. It is only since the rise of the industrial research laboratory in the past 50 years that science has been carried on to any appreciable extent in government or in industry, or by the professional research worker. Also, in the past scientific advice was needed by governments only rarely, and governments were perfectly happy either to do without such advice, or to turn to national academies or other non-governmental bodies. In fact, govern ments in the past, and to some extent still, have been rather suspicious of science because the laws of nature cannot be established by the customary legislative process, and are in no way susceptible to government policy.

With the increasing importance of science to technology, and incidentally to defence, science has relatively recently and suddenly begun to impinge on government policy. Under these circumstances the need for government scientific facilities and permanent advisers has arisen, and there has been an enormous increase in government science. The fact that science has become a matter of government policy and interest has brought with it the customary advantages and disadvantages. Public interest always means increased support, but also always means increased control and diminished freedom. The reconciliation of these two opposing factors in government, in industry, and even in universities, is the most important problem facing science in the next decade or two.

As science has become necessary to government, two distinct types of organization have developed. First there are many government departments that need experimental facilities in order to do their jobs. In Canada these include Agriculture and Mines among others. Both these departments have Acts of Parliament to administer, and both also have responsibilities (to the farmers or the miners) for scientific work in their fields. In a sense, but fortunately only in a very broad sense, their scientific work is circumscribed by the terms of reference of their departments, and their work has a definite applied objective. Many countries have set up a different type of body, in Canada represented by the National Research Council, whose duties are rather like those of a national academy, and include the support and encouragement of work in pure and applied science, together with a residual responsibility for scientific research in all fields and especially in those not covered by the more narrowly defined objectives of government departments. Such bodies need, and often have been fortunate enough to obtain much more freedom than a government department. They are very complex organizations intimately connected simultaneously with government, with industry, and with universities. Canada has been somewhat unusually enlightened along these lines, and I would like to make a few remarks about the National Research Council.

The National Research Council was set up by Act of Parliament in 1916 and is a corporate body, not a department of government. It has no Minister in the usual sense, but reports to a Committee of the Privy Council on Scientific and Industrial Research, which is composed of nine Ministers whose departments have to do with research or scientific affairs. The Chairman of the Committee has for a considerable time been the Minister of Trade and Commerce. It should be emphasized, however, that the Council has nothing to do with the Department of Trade and Commerce as such. It is, in fact, one of the first examples of a Crown corporation. The Act gives the Council a number of powers which are not possessed by government departments: in particular, the Council is outside the Civil Service, has a governing body of independent, non-government scientists, can earn revenue and spend it, can build its own buildings, and so on. All these powers have been used with discretion, but they are absolutely essential for the operation of a first-rate scientific organization with broad responsibilities. Above all, it is vital that the high reputation of the scientific staff be maintained. The status of the staff is due entirely to the control being in the hands of the Advisory Council, a group of the most distinguished scientists in Canada. We are, in fact, one of the few countries which has recognized the fundamental fact that the control of a scientific organization must be in the hands of scientists. At first sight this does not appear to be a very great accomplishment. It would appear to be obvious that an organization should be controlled by people who understand what it is for. This is, however, extraordinarily rare in modern society. In fact, it is almost a principle of modern administration that the control of an organization should be in the hands of those who do not understand it. It is, therefore, a major accomplishment of the Canadian government that in many cases our scientific and other specialized agencies are controlled by people familiar with their purposes. Certainly in science we are much envied by many foreign laboratories on this account, and Canada has had a great deal of influence on the organization of many foreign scientific bodies.

The general functions of the Council are threefold: to advise the government on matters of science in general; to encourage and promote fundamental science, especially in the universities; and to operate laboratories of its own, especially in fields which are not the responsibility of government departments, and with reference to the problems of secondary industry.

The general organization is complex and very much decentralized. Because governments tend to centralization this presents many problems. Decentralization has been accomplished because of two main factors. The first of these is the development of an administration whose main function is the protection of the scientist to the greatest possible extent from the red tape inevitable in government operations. The second factor has been the far-sighted attitudes of successive governments in leaving the Council free of many hampering restrictions. In every field the problems differ and each scientific division thus has special features of its own. Nothing could be worse than the "Big Organization" point of view which regards uniformity of administration as an end in itself. Financial control is, of course, essential. The balance sheet, however, is no more the criterion of efficiency for a scientific organization than it is for a charitable one. In each case the money must be well spent, but the real results are rather intangible.

From the experience of the National Research Council, there is a wide variety of types of work which such organizations are called upon to do. As far as the National Research Council is concerned the list includes fundamental work, long-term applied work with no specific objective, work on specific industrial problems, short-term industrial problems (i.e., *ad hoc* investigations), investigations for the Services, consulting, testing, specifications, and miscellaneous enquiries. All of these are of importance, but it is essential if the organization is to develop any reputation or scientific self-respect that the *ad hoc* problems and routine enquiries shall not be allowed to force real research out of the door. It is very easy for this to happen and in the case of many laboratories of similar type in other countries it has happened. In my view, at least as far as the National Research Council is concerned, long-term investigations, fundamental or applied, *must* constitute the major effort of the laboratories if they are to keep the scientific reputation they have earned. . . . The National Research Council has three separate reputations to consider: with the government if it is to get support; with industry if it is to be useful; and with universities if it is to have any scientific reputation, to have competent staff, or to be of any use in the promotion of Canadian science.

The Council has, I think, acquired a unique reputation over the years. This has been due to three things: first, it has a governing body which is composed of non-government people, but these people are almost all eminent scientists; second, it has enjoyed far-sighted treatment from the governments of the day which has left it free from many of the normal aspects of government control and interference; finally, the most vital feature has been its dual function. By being responsible both for university support and for the operation of laboratories it has been able to avoid two pitfalls. In the first place, as an operating laboratory it can retain first-rate scientists, and avoid a narrow bureaucratic outlook in dealing with the university programme. Secondly, by maintaining the interest and active participation of university people it can maintain the scientific standards of its laboratories.

CHEMICAL INSTITUTE OF CANADA
MONTREAL, OCTOBER 17, 1956

. . . It should be emphasized that there is no real clash between the interests of the private consultant and the government laboratory. In general, the government laboratory is interested in keeping to problems as long-range as possible. It is a general rule not to undertake anything which can be done effectively by a private laboratory. . . .

SPECIAL COMMITTEE ON RESEARCH: 24TH PARLIAMENT
JUNE 2, 1960

. . . The fundamental feature of the administration of the Research Council . . . is to make sure that the administration can never issue any instructions to scientists in connection with any technical subject whatever. This is the fundamental principle of our administration. It is the exact opposite of the administration of most government departments, where the administrative head is in charge. In fact, the scientific divisions have responsibilities to the senior director, who is an active scientist, and the engineering to the vice-president (scientific), who is an active engineer; and all divisions report directly to me on scientific matters whenever they feel like it. It is up to the administration, which also reports to me, to make sure that things can be worked out with the scientific divisions, so that the administration act as a service to the divisions, rather than as a control. The result is a highly decentralized organization.

I think that the organization is almost unique from the point of view of scientific organizations, and I might say that almost every government research laboratory in the world is trying to copy it—in some cases successfully, and in other cases, unsuccessfully.

But the main principle is that you do not allow the scientific programme to be directed by the administrative group, and I think this is the most fundamental principle of the whole system. . . .

SASKATCHEWAN RESEARCH COUNCIL
OCTOBER 1, 1958

. . . The most striking differences between Britain, the United States, and ourselves occur in our attitudes towards government laboratories. In the United States the industry is strong and perhaps might be said to be a little suspicious of the government laboratory. In view of the magnitude of the American industrial research effort, the government contribution is usually no more than a certain degree of reinforcement, and is often at a rather basic level. Exceptions are those industries, such as the electronics and aviation industries, which are financed largely by defence contracts. The research association, where a group of industries in the same field pool their research effort, has never been a popular American device.

In Britain it is often, though perhaps not quite accurately, stated that fundamental work is well in advance of applied. As a result there is a tendency to worry about the application of existing knowledge to industry, and in my own personal view to over-organize industrial research. The result is that research associations have been successful in a number of cases. There is often a tendency on the part of industry to look to the Department of Scientific and Industrial Research for guidance in a way which would be inconceivable in the United States. As a result the problems of a British government research organization like the Department of Scientific and Industrial Research are quite different from those of the National Bureau of Standards or other United States government organizations.

In Canada the problems are quite different from those of the Department of Scientific and Industrial Research. Research associations are feasible only where the Canadian industry is large from an international standpoint. As a result the only major association of this kind is the Pulp and Paper Research Institute in Montreal. In general, however, the research association has not been popular. There is no use asking a branch company to join such an organization, because it is already, with subsidiaries in other countries, a member of a parent research and development group. . . .

In Canada we have no large semi-public applied science institutions such as the Mellon Institute, Battelle Institute, Stanford Research Institute, and so on. We also lack large private industrial laboratories such as those operated by companies like DuPont, General Electric, Shell Oil, Bell Telephone, and others. Furthermore, considering our size it is not likely that we will have such institutions in the foreseeable future. It therefore seems essential that such large-scale laboratories be maintained and operated by the federal and provincial governments. As a matter of fact, the main difficulty with institutions like Mellon, Battelle, and Stanford is that their income has to be earned to too great an extent, and they therefore swing far too much towards *ad hoc* problems. Such laboratories need to be free to follow their own inclinations. This means that the job can be done much better by an institution with government support. No institution can do research of high quality unless it can devote a considerable share of its effort to problems of its own choosing.

There is another factor of importance to Canada. Though in the United States the comprehensive research laboratories of the large industrial corporations serve directly only a relatively few industries, these laboratories do exercise an indirect influence over applied scientific thought across the country. Even the technical sales and service laboratories of such corporations often act as technical advisers to their customers. The lack of large research laboratories in Canadian industry puts their technical services one remove farther away, and makes government services such as the Technical Information Service still more essential. This is a field in which local organizations such as provincial research councils can play a major role. The system of agricultural organization is an excellent illustration of what is needed. This goes from fundamental and applied research, through experimental and demonstration farms, to educational and extension services to individual farmers. There has never been any doubt that it was the responsibility of the state to serve the individual farmer, or that he was incapable of operating his own private research laboratory. The small industry occupies a position somewhat similar to that of the individual farmer, and its technical assistance is equally essential. This is certainly a major task of a government research institution, and especially for a provincial institution. . . .

CHEMICAL INSTITUTE OF CANADA
MONTREAL, OCTOBER 17, 1956

. . . There is one great paradox about the effect of the war on Canadian science. It is a truism that during a war no real science is carried on. The people who would otherwise be doing research tend to be engaged in a variety of practical problems, vitally important for war but in no real sense constituting research. This was as true in Canada as elsewhere, and yet in a real sense Canadian science came of age during the war. There were several reasons for this. In the first place, our traditional isolation was broken, and Canadians began to see something of their colleagues elsewhere, and in the process began to realize that because an effort was Canadian it was not necessarily second-rate. In the second place, the war provided facilities which had been sadly lacking, especially in the universities. Finally, the war provided recognition for good men and good work. The result was that, although scientific work in any real sense came to a stop during the war, nevertheless, at the end of the war Canadian science was in a position to come into its own. The important thing, however, is to realize that there is still a long way to go: that we have now grown up sufficiently that we must recognize second-rate work as second-rate, and give up the idea that such work may be "good by Canadian standards." This means that we have no cause to be complacent, and that there are many fields in which our work is not as good as it ought to be. It also means, however, that there are many other fields in which we can view Canadian work with some real satisfaction. . . .

The work of the National Research Council laboratories is divided up into divisions, based on fields of science. We have always strongly opposed organization by projects, because we feel that the only way to develop first-rate people is to keep them working in their own field. Essential projects are, therefore, fitted into the divisional structure, responsibility being given to the division to which the project is most closely related scientifically. . . .

There are many reasons which justify fundamental work

in a laboratory like N.R.C. In the first place, there is every reason to make a contribution to Canada's position in fundamental science. If the work is of high quality it reflects favourably on Canada's general scientific standing, and on the reputation of N.R.C., especially internationally and as far as academic institutions are concerned. This has a highly beneficial influence on our ability to attract first-rate men, and brings to the laboratory on fellowships people from all over the world. Finally, the existence of a fundamental group for consultation and discussion has an extremely helpful influence on the outlook of those doing applied work, and on the general atmosphere of the laboratory. . . .

CANADIAN INDUSTRIES LIMITED
BELŒIL, OCTOBER 22, 1954

. . . In a research organization a few people make all the difference. If 5 per cent of the staff of a research laboratory are really first-rate, with imagination and initiative, all is well. Without this 5 per cent very little that is worth while will emerge from the laboratory. If we are going to expand, these are the essential people. This is where the shortage really is and it is where the acute shortage will develop.

The problem is to develop people of this type: to get behind them when they appear and give them the opportunity to develop themselves. In both Britain and the United States the source of supply of such people is from the universities. The problem is how to hold more people of this type in the universities where they will help out the training of research students, and themselves develop the experience and ability to direct research. . . .

ACOUSTICAL SOCIETY OF AMERICA
OTTAWA, MAY 15, 1959

. . . The problem of organizing science, government or other, is a difficult one, and one which is particularly under discussion these days. Above all, if scientific work is to be creative and effective we must avoid its over-organization and the efforts of so-called efficiency experts. It is good occasionally to remember the definition of a camel: a greyhound put together by a committee. One must also try to avoid the stifling effects of bigness. The tidy mind whose ideal is "one big organization" is the worst foe of originality, initiative, and scientific progress. Some uniformity is, of course, necessary for administrative reasons, but let us at least avoid conformity for its own sake and regard it as an evil which is unfortunately sometimes necessary. The 64-dollar question, of course, is how much organization is needed. . . .

In general, there is no question that too much organization and planning is far more frequent than too little. I have read a number of books on the organization of research, and they would be hilariously funny if the situation were not so serious. There is no question that the only reason that industrial (and government) research is reasonably effective is that the attempts of the planners and organizers have not been completely successful.

All the arguments about over-organization of research in industry or universities apply *a fortiori* to government, because governments are always very highly centralized, especially as regards personnel policies and financial control. Industry has frequently recognized, by setting up subsidiary companies, that the same administrative methods cannot be applied to manufacturing plants and to research laboratories. There is no doubt that the organization of science within a government framework is very difficult, and many governments in certain cases have used devices comparable to the research subsidiary in industry. . . .

INDUSTRIAL RESEARCH INSTITUTE, INC.
QUEBEC CITY, OCTOBER 27, 1952

. . . Finally, I would like to make a few comments on some of the main problems involved in any research organization.

Youth

The great advantage of a university is the way in which the average age of the group remains almost constant, no matter how old the staff is. A continual influx of young men is essential to any organization that wishes to stay alive. It is, however, difficult to maintain, and there is always the danger of the organization resembling an old people's home after a few years. In industrial laboratories the drain to administrative and operating positions may be enough to keep positions open for new blood. In a government laboratory, where there is no operating side, the position is much more serious. One solution which we have used for work on the fundamental side has been to maintain a flow of post-doctorate fellows. The whole problem, however, is one which requires constant attention.

Secrecy

Under our present conditions there are many cases in which security restrictions are essential. The danger, however, is that it is more and more being taken for granted. There is some danger of a generation growing up who do not see anything intrinsically wrong in security restrictions. The main thing which must be kept in mind through a period such as the present is that, although secrecy may sometimes be necessary in scientific work, it is nevertheless intrinsically deplorable.

Co-ordination of Research

Finally, I would like to offer a few comments on a much-discussed subject, the co-ordination of research. In the wake of the effort of the last war, and of large military spending, we are seeing a fundamental change in outlook towards research. Vast "co-ordinated" efforts are being made, with the inevitable attendant inefficiency. Even the fundamental workers at universities are being steadily sucked into this. The result has been a fundamental change in the philosophy of research laboratories.

Planning and exact co-ordination are, of course, strictly necessary in a specific programme once it has reached the development level. The question is whether such ideas can be carried over to research (pure or applied) which has any long-term significance. The whole question is a burning one at the moment, and I have no solution to offer, but I should like to make a few rather disjointed remarks. First, ideas are really what are wanted. A compact streamlined organization which follows a nice chart and is co-ordinated at a high level will never produce an original idea. The inevitable result of a strictly planned research programme is (i) a director who is probably so senior that he is entirely divorced from technical reality, and (ii) a group of people who are reduced to the level of technicians. Secondly, I feel strongly that the inefficiency of large co-ordinated programmes is something which is very hard to combat, except by extreme decentralization, which in effect means sharply decreasing the amount of planning. Unfortunately, the empire-builders are always with us, and any such decentralization will usually meet with much opposition. Finally, the larger the organization the greater the predilection for pieces of paper. Essentially this means more and more "progress reports," usually reporting little progress. The idea, of course, is that everybody is "kept in touch" with everyone else. I wonder if it wouldn't be better to follow normal scientific practice and report only when there is something to say. . . .

Planning is certainly a necessary thing, and so is co-ordination. Co-ordination for its own sake, however, can kill initiative faster than any other factor. I think planning, like security, should be regarded as an evil, even if it is an essential evil.

MANAGEMENT CONFERENCE, QUEEN'S UNIVERSITY
JUNE 16, 1959

. . . All these problems lead to the real major difficulties, evident to those engaged in scientific work, but not to the public or, I'm afraid, to many of those responsible for the administration of major scientific organizations. These major problems are: How should science be organized? How important is team-work? What is the status of the individual in science?

To what extent should the scientific efforts of a laboratory be organized? A certain amount of organization is, of course, necessary. There must be light, water, and power, people must be paid, there must be technicians and workshops, the better people must have assistants, the less experienced people must have some guidance, and so on. I am certainly not advocating anarchy, but the problem is that creativeness and organization run entirely counter to each other. This is why the real efficiency of a university laboratory is vastly greater than that of any industrial or government laboratory. The important thing is that in any well run laboratory there must be a conscious and continuing effort to reduce organization and planning to a minimum, to have as few committees as possible, to write reports as infrequently as possible, and to regard "co-ordination" as a dirty word!

. . . Ideally, because pure science has no specific end in view, there should be as close to no organization and planning as is humanly possible. Because development has specific aims, however, a development programme should necessarily be highly organized. Even here, however, one must make sure that organizing is not being carried on for its own sake, and that accountancy is not the dominant influence in the laboratory.

Probably the worst thing that can be done is to bring in a firm of management consultants, who know nothing of science, and ask them to organize you. If this is done you can be sure that the work of the laboratory will be lousy! The most striking example of what can be done in this way is illustrated by a recent speech in which a management specialist said that you did not have to understand an operation in order to make it efficient. I must say that I feel strongly that the only reason why industrial research is reasonably efficient is that the steady efforts to over-organize it have providentially been relatively unsuccessful so far. I also must object to the main principle of modern administration that the control of any operation must be in the hands of those who do not understand it.

It should be mentioned that there is a great deal of concealed and indirect planning of pure science. Thus most nuclear scientists probably think they are engaged in pure science. It is, however, only because of bombs and power that large accelerators costing 50 to 100 million dollars are available. The result is that government support essentially has converted nuclear physics into long-range applied science. This is also true of upper layer atmosphere physics, oceanography, and many other fields. The resulting "planning" of pure science is going to cut down on the possibilities of really new discoveries, because planned pure science will always follow the path which is obvious at the moment. Of course, financial control is essential when very large expenditures are involved, but its disadvantages should be realized, and there should not be too much effort to co-ordinate such planning.

Another facet of the situation is the status of the individual as compared with that of the team. There is a popular view that team-work is the modern way to do research, and that the day of individual accomplishment is past. I don't believe a word of it. A team has never had an idea and never will. What it will do is drive relentlessly on towards the obvious conclusion.

In support of these views I would like to quote from two people who are not scientists. The first is John Stuart Mill, who remarked that the question is "whether our march of intellect be not rather a march towards doings without intellect, and supplying our deficiency of giants by the united efforts of a constantly increasing multitude of dwarfs." On the basis of Mill's remarks, the ultimate end of the team is the computing machine. These are often called "electronic brains," which of course they are not. They are merely capable of an endless and rapid repetition of simple operations. The electronic computer is, in fact, the equivalent of a team of millions of docile morons. It is a highly useful device, but I'm afraid that it has many of the qualities of a large committee.

Of course teams have always existed. In the old days in science they consisted of a university professor and a few students, and in my mind this is the ideal team. The trouble with a large team, however, is that it inevitably leads to over-organization and over-planning. If this is carried too far one man is making all the decisions and the others are merely pairs of hands. Also, the man who *is* making the decisions is too busy organizing to be able to think about science.

A second quotation is from Jewkes.

> It seems to be possible to exaggerate the virtues of team-work. . . . Quantity cannot make up for quality. The reasons for the

limitation of team-work are obvious. Team-work is always a second best. There is no kind of organized, or even voluntary, co-ordination which approaches in effectiveness the synthesizing which goes on in one human mind. . . . A large team is essentially a committee and thereby suffers from the habit common to all committees but especially harmful where research is concerned, of brushing aside hunches and intuition in favour of ideas that can be more systematically articulated.

In so far as society can usefully interfere, its task might well be to try to maintain the balance between the different sources of invention, to strive to prevent any one dominating to the exclusion of others. That country will, therefore, be happily placed which has a multiplicity of types of research agencies. . . . As contrasted with the ideal ways of organizing effort in other fields, what is needed for maximizing the flow of ideas is plenty of overlapping, healthy duplication of efforts, lots of the so-called wastes of competition and all the vigorous untidiness so foreign to the planners who like to be sure of the future.[1]

Another feature of the team is a reverence for data in quantity. It is often said that so many data are being collected in some fields that it is impossible to work them over or make use of them. This would be the danger of extending the International Geophysical Year for too long, and substituting too much data-collection for original thought. The obvious answer is why not collect fewer data and think more. The difficulty is that it seems to be psychologically impossible to turn off an automatic device. There is such a thing as too much co-operation in science, because co-operation inevitably diminishes initiative.

Where are We Going in the Future?

Finally, one might ask, where is science going, and how much of it do we want? At the present time in some highly industrialized countries about half of 1 per cent of the population are engineers or scientists. Of course, most of them are

[1]John Jewkes, David Sawers, and Richard Stillerman, *The Sources of Invention* (London: Macmillan, 1958), pp. 162–63.

not engaged in research, but about a quarter of them are. The rate of expansion of research is about twice that of G.N.P. or a little more. Research not only is expanding, but the *rate* of expansion is also increasing. Obviously this cannot go on forever. At the present rate, in a century or two everyone will be engaged in research and no one will be left to provide food, clothing, or services. Some time we must level off, but when? Certainly not for a few decades yet, because the increase in technological innovation is expanding production rapidly.

How many scientists do we need? The popular outcry sometimes suggests that the number is unlimited, but obviously a slow-down will come, though not for some time yet. It is worth contrasting the present position in science with the situation for education in general in the Middle Ages. As society got more complex, more and more people were needed who could read and write and had some rudimentary education. In the Middle Ages such an education was possessed only by university graduates. As a result there was a big expansion in universities to provide the civil servants and administrators of the day. We now need this much education, or more, from a junior clerk, and elementary schools supply it.

Isn't most so-called science and engineering today really a job for someone with a general education which includes some awareness of science and its methods? In short, haven't the humanists really sold out to the scientists as far as many careers are concerned by refusing to include any appreciation of science in a general education? If in our science-based society enough appreciation of science were included in a "general education" we could probably do without half the engineers and scientists needed today. In my view, in the future we will pass over a hump and begin to have a diminishing percentage of engineers and scientists in our society, but they will be better engineers and scientists and they will really do engineering and science.

The Civil Service Commission and the Outside Agencies

PREPARED FOR MEETING OF THE INSTITUTE OF PUBLIC ADMINISTRATION
1959

THE CIVIL SERVICE COMMISSIONERS concluded their recent report, "Personnel Administration in the Public Service," with two short paragraphs which say, in brief: first, that Canada has been fortunate in having a good Civil Service over the past several decades; and second, that if legislation is enacted along the lines of the proposals contained in the report, things will be still better than they are.

We agree wholeheartedly with the first statement. Anyone who has observed the Civil Service of other countries in action cannot feel other than proud of the Canadian Service. There is no doubt that a high degree of competency has been achieved, and that the future will see further desirable developments. As far as the National Research Council is concerned we compliment the Commission on their accomplishments, we wish them well in the future, and we are anxious to help in any way short of endangering our own effectiveness. However, on the second statement I strongly object to the recommendations for outside agencies.

It seems to me, as one concerned with the scientific outside agencies, that I should first point out that the report of the Civil Service Commission really consists of two separate documents which, at least philosophically, are mutually incompatible. The major part of the report expresses the general desire for more decentralization and for more freedom of action on the part of departments. The second part, Appendix C dealing with the outside agencies, suggests more centralization and far less freedom of action on the part of these agencies. In other words, the report is essentially suggesting an averaging process whereby one vast uniform system will be produced by making some groups more free and others less.

I would like to emphasize that my arguments are directed to the report and not to the Civil Service Commission as such. I certainly have no desire to set myself in opposition to the Commission nor have I any desire to suggest that the Commission, within its own defined functions, is not doing a good job and is not hoping to do a still better one. I am going to confine my remarks to the question of the report as it deals with the outside agencies and here my views on it are completely adverse.

The report covers the whole of the Public Service, but it has been prepared by a body responsible for only a part of the Service, and without reference to the views or experience of the agencies outside the Service. In other words, it is not a judgment but rather a brief for the prosecution, and I wish to put in a brief for the defence. The body of the report draws on the vast knowledge, experience, and competence of the Commission but Appendix C has no similar basis of experience and understanding. It may be a little unkind and a slight exaggeration, but it is not unfair to suggest that the body of the report is a professional document, while Appendix C is an amateur one at least as far as scientific organizations are concerned.

In order to set the problem up for discussion I would like to quote from two separate sources which take a diametrically opposite point of view. The first is an editorial in *The Ottawa Journal* of February 7, 1959, entitled "The Research Council and the CS Commission." It runs as follows:

The recent report on personnel administration in the public service, presented to the Government by the Civil Service Commission, has been several times welcomed and praised in these columns. It is in the main a thoughtful analysis and even such specific recommendations as one might not agree with are phrased in a tentative way that can invite only discussion and not abuse.

One such recommendation which *The Journal* would like to see thoroughly examined before its being approved is the advice of the Heeney report that the National Research Council be placed under the Civil Service Commission.

Since its inception the Research Council has been exempt from the provisions of the Civil Service Act, and there is some reason to believe that this very exemption was a contributing factor to the Council's rapid and impressive attainment of high international repute. It has been able to lure to itself some of the brightest and most adventuring minds, Canadian and foreign. These "types" do not take easily to regimentation, they cherish irregular hours and bridle against classification; security is not so much their conception of a good job as is the opportunity to think and probe and experiment. Moreover, the standard principles of promotion *via* seniority could conceivably ground a scientific research team in short order. (Already there is regret at the number of good scientists leaving Canada for the United States—would not the proposed change to the Research Council increase that number?)

In stating these thoughts we do not suggest it necessarily follows that departments under the Civil Service are hopelessly bound and made inefficient by the Commission's requirements. We simply believe that all departments do not call for precisely the same rules of appointment, promotion, and work conditions. Indeed, the Civil Service Commission report itself admits this for it would continue to exempt from the Commission's provisions such bodies as Atomic Energy of Canada, Bank of Canada, Trans-Canada Air Lines and many others.

The Heeney report anticipates argument on this matter by saying, among other things, that if science and research in agriculture, for instance, can be conducted under Commission conditions in the Agriculture Department then presumably the Research Council could do the same. We will not take time to discuss that reasoning other than to say we find it singularly unimpressive.

The report further defends this recommendation by saying that its other recommendations are going so to improve the Civil Service Act's provisions that such as the Research Council would find their objections to the Commission removed. To that we must reply that it is an "iffy" proposition. If the Commission were greatly improved then the Council *might* not be harmed; but if it is not improved sufficiently or sufficiently fast then the Council could be greatly harmed. One is tempted to say to the Commission that it should first try to put through its proposed improvements and then see whether they are of a nature sufficient to safeguard the peculiar characteristics of such as the Research Council.

Finally, *The Journal* would say this: we are greatly in favour of most of the Commission's recommendations, and certainly the improvement and bringing up to date of the provisions of the Civil Service Act is a most desirable and even urgent goal. But we must be wary of the tidy mind. It sounds fine to assert that almost everything in public service should be brought under one surveillance, it has a ring of efficiency to it. But it is a generalization to be examined with care and even suspicion.

The second quotation is an excerpt from an address by one of the Civil Service Commissioners to the 39th Annual Meeting of the Professional Institute of the Public Service of Canada. I am going to address most of my remarks specifically to this quotation. This is in no sense because I wish to take issue with the Commissioner rather than with the report, but because the remarks are an admirably concise summary of the spirit of Appendix C, and contain in a short space all the ideas to which I am implacably opposed. The quotation follows: "Finally, I would like to mention our recommendation for 'one Civil Service'. To an audience such as this I do not need to dwell on the dissatisfaction that has been expressed about the treatment received by, say, a scientist employed in the Civil Service and by his counterpart in the various exempt agencies. Since all are employees of the same Government, it seemed to us that policies and procedures of personnel administration should be the same for those with similar qualifications and comparable duties and responsibilities, regardless of where they are employed in the Public Service."

I think one can find the cause of disagreement readily in this address. In the first place, are there many discrepancies between the Civil Service and the outside agencies? In view of the fact that both the outside agencies and those within the Service are ultimately dependent on Treasury Board control, I feel that there is far closer co-ordination than the report suggests. Secondly, suppose that there are discrepancies; the question then arises as to which side is right, and I see no excuse for suggesting that the faults all lie on one side, at least not without a serious examination of the situation. The third point is: is it necessarily undesirable to have discrepancies?

It should be pointed out here, however, that there is a considerable difference in outlook between the National Research Council and the Commission on organization. This arises from the fact that the Commission is engaged in general with the staffing of large administrative units of government while the National Research Council is engaged solely with the operation of scientific laboratories and the attempt to produce creative work under government auspices. Creative work, of course, is also carried on in orthodox departments of government. It is, however, a relatively minor part of the Commission's responsibilities while it constitutes the whole of the responsibilities of the National Research Council. This leads to many fundamental differences in viewpoint.

The major difference involves the questions of organization, classification, and so on. When dealing with administrative matters in general it is obvious that job classification and organization are of major importance. In a scientific laboratory they are not only of minor importance but in many cases are of no importance whatever. For example, whenever the government sets up a new office, the organization required is drawn up, the duties of the positions determined, and the qualifications for each job written. The distinctive point is that the job determines the salary. Finally, the jobs are advertised and the applicants best qualified appointed. On the other hand, in a research laboratory scientists have only one job to do and that is to do good scientific research and the better they do it the more they should be paid. In consequence the fundamental policy is to write the organization chart after you have employed the staff and to ignore it totally beforehand. This leads to discrepancies but they are highly desirable discrepancies, and if relative salaries of the Council's employees and those in the Civil Service are examined from this point of view it will be found that most of the so-called discrepancies disappear.

I feel that because of the fact that the organization of the Council is in the hands of working scientists themselves we have been much more successful in making clear to the Treasury what the real needs are in the way of senior professional salaries. I am, of course, not in any way suggesting that we have no organization at all. We have, in fact, one of the smoothest running organizations in the Public Service. The distinctive point is, however, that we fit the organization to the man, not the man to the organization.

Further, it should be emphasized that in the operation of a scientific laboratory nothing matters much expect the competence of the staff. No man given the responsibility for maintaining a first-class research laboratory can afford to relinquish the responsibility for its personnel to a second person or to an independent body that is not responsible to him and whose interests are not directly linked with the success of his laboratory. There is no compromise on this view as far as the scientists are concerned; and there are plenty of examples throughout the world to prove that the principle must be followed if an absolutely first-rate, top-ranking scientific laboratory is to be maintained.

The second point lies in the emphasizing of the fact that the Council is part of the Public Service. In our view this is one of the least important things about us. The Council has rather complex responsibilities. We are, in effect, five things at the same time. The first is a government laboratory with certain narrowly defined specific duties: this is a minor part of our work. Secondly, we are in many respects a foundation with purposes almost identical with those of the Canada Council. In the third place, we are an industrial research laboratory similar in many ways to places like the Mellon Institute, or to the laboratories of major industrial firms. Fourth, we are a research institution much more like the Rockefeller Institute or a university laboratory than a government department, and finally, we have many of the functions of a national academy, functions similar to those exercised in Britain by the Royal Society of London or in the United States by the National Academy of Sciences.

If, then, we are not allowed to be ourselves but must resemble someone, and must have administrative uniformity, whom should we pick? Are we to be organized like a national academy, like a university laboratory, like the Canada Council, or like the Department of Public Works? My own feeling is it is obvious that if we are to resemble anything we should resemble something that is functionally similar and therefore should choose an academic research institution as a model. What we have striven for is to be as similar to a university laboratory and as unlike a government department as it is possible for us to be and still carry out all our functions successfully. The result has, I think, been quite successful. We have earned the confidence of the government as a government scientific laboratory and as scientific advisers, while maintaining the respect of the universities as a Foundation, the respect of industry as an industrial laboratory, the respect of the major research institutions as one of the world's great research laboratories, and have kept a close and sympathetic relationship with national academies in most of the countries of the world.

It should be recognized that the operation of scientific laboratories of large size by industry or by government is a relatively modern phenomenon. Such laboratories did not exist to any appreciable degree 30 or 40 years ago. It is noteworthy that since the time of this development every major scientific research organization set up by the Government of Canada has been put outside the Civil Service. When the Government of Canada revised the National Research Council Act in 1924 to enable it to operate laboratories, it understood the importance of establishing and maintaining an academic atmosphere for scientific creativity. For this reason it placed the Council outside the Civil Service, and on several occasions since then various governments have reaffirmed the original decision.

Actually Canada was a pioneer in government laboratory organization. The Council's Act has been copied by many countries all over the world. Recently, Edward McCrensky, Director, Civilian Personnel and Services Division, Office of Naval Research, Washington, D.C., had this to say in his excellent book, *Scientific Manpower in Europe*: "European experience largely confirms the importance of leaving the selection and promotion of scientists, so far as possible, in their own professional hands." It is ironical that, in 1959, in Canada, the country where the idea of scientific independence was fostered and where its soundness has been proved so successfully, a recommendation should be made to destroy it.

The National Research Council Act placed the responsibility for the selection and supervision of the Council's personnel in the hands of an honorary advisory body composed of some of the most distinguished scientists in Canada. As a safeguard both the President of the Council and the Minister were given the right to veto the actions of the Advisory Council, in that the nomination of the President and the approval of the Minister must be obtained before an appointment or promotion is made. The important fact is that under the National Research Council Act, only the Honorary Advisory Council of scientists has the right to appoint or promote anywhere within the organization. This body has the confidence and respect of the staff which a purely non-scientific body could never hope to attain. It is of prime importance to maintain a high level of morale in a creative staff and this cannot be done by placing the control of their personnel policies in the hands of an organization completely outside their own professional field or by trying to fit them into a uniform civil service organization.

There is no question that creative work runs counter to administrative organization. Organization is necessary but it must be sympathetic and it must be reduced to a minimum. The worst possible attitude for an organization which is attempting to do creative work is an insistence on uniformity as a good thing in itself. In a major scientific institution the main thing is to develop a character and an atmosphere which distinguish the organization from all others. In my view everything that can possibly be done to make such an organization different administratively should be done. I cannot emphasize too strongly that I feel that the most undesirable thing in any creative organization is uniformity. Some uniformity is, of course, necessary but it should be regarded as an unavoidable evil rather than as a desired goal.

One of the great troubles with the development of a more complex society is a steady move towards larger and larger units, both governmental and industrial, and thence to more and more complex administration. To my mind there is no similarity whatever between the proper organization of large administrative departments of Government and the organization of a laboratory or other body designed to carry on original work. This fact is becoming recognized everywhere and has certainly largely been recognized by industry. With more and more frequency industries are setting up subsidiary companies to carry out their research for the one and simple reason that they find it impossible to do research under the same administrative set-up required to operate plants.

The same trend is evident in government research throughout the world. In fact, the National Research Council has served as a model for government laboratories in many Commonwealth and foreign countries. Similarly, because of our reputation and experience, we have been more and more frequently consulted by universities, by industries, by provincial governments, and by foreign governments in order that they can benefit from our knowledge of scientific organization. It is not too much to say that the National Research Council, the Defence Research Board, and Atomic Energy of Canada Limited have become much admired patterns for the organization of government laboratories in many parts of the world.

Finally, it should be pointed out that with the experience I have outlined we obviously are far more qualified to decide on questions of scientific organization than could possibly be the case with any body which is not actively pursuing scientific work itself. Under the circumstance there is no question that it would be a thoroughly unfortunate step to move the clock back 30 or 40 years and to overlook the high reputation which has been obtained by many Canadian government scientific agencies because they have been able to preserve in their government organizations the traditional freedom of science in the universities. I feel that a number of outside agencies can serve as models towards which the Commission can strive in its sincere and whole-hearted attempt to improve the flexibility of the government Service. I certainly do not feel that this objective can be obtained by first knocking us down and then trying to build from the ruins a comparable structure for the Service as a whole. Again, I would emphasize that above all I refuse to submit to the view that uniformity is a good thing in itself whether it be administrative or otherwise. If uniformity is the goal of progress the future of mankind looks unbelievably dismal. . . .

The Organization of Science in the Soviet Union

CANADIAN PUBLIC HEALTH ASSOCIATION
OTTAWA, DECEMBER 1, 1960

. . . IN COMING to the discussion of Soviet science it is most important that we differentiate clearly between science and engineering and technology. As pointed out above we *are* engaged in a race with Russia and questions of prestige, publicity, and so on cannot be ignored. The important thing is to recognize such factors for what they are and to avoid allowing the "race" outlook to destroy the integrity and the effectiveness of science in the free world.

In dealing briefly with Soviet science I would like to keep the discussion on the philosophical rather than on the travelogue basis. To a considerable degree the impressions received by any visitor will depend on his preconceived ideas and will differ from those of any other visitor. Also, the impressions of science in any country will be dependent largely on the status of the visitor's own special field. I do not think, therefore, that the views of one more physical chemist on work in his own field are particularly enlightening. I knew before I visited Russia that work in my own field was good, and my opinions were confirmed. I think, however, that it is much more interesting to discuss what we might expect from the *system* in the U.S.S.R. as it concerns science.

When a visitor returns from the Soviet Union there is one question which is asked over and over again, which becomes rather irritating, and which is completely unanswerable in a direct way: Where do we stand in the race, or, is Russian science ahead of the West? Science, of course, is made up of a large number of distinct disciplines, and each of these can be broken down into many fields. In each field the progress in research in a given country depends on the ability of a few persons, or even of a single person. In a given field there are often only seven or eight major laboratories and any country, big or small, may have none of them, or even all of them. It is thus relatively easy to assess the status of a given country in a small field. It is a bit dubious in a larger field, silly in a whole subject, and quite meaningless for science as a whole. It seems to me that the only sensible answer that can be made is that the Soviet Union is in the Big League. That, like the United States or the United Kingdom, it will be good in some fields and poor in others, *but* that there is no reason to assume that it may not be at the top in any given field. In short Russian science has come of age, and if we insist on regarding science as a race we must not delude ourselves about its position.

There is no question today that science in any country is now intimately bound up with economic and other factors and as a result priorities in science will be warped by factors of national importance. The more the over-all economy is planned the more we may expect such a warping of scientific priorities on the basis of end use. In the long run such warping of the structure of science will be inefficient for no one is wise enough to predict what will be important in 10 or 20 years. Because the structure of scientific organization is much more monolithic in the U.S.S.R. than in the West we may expect that such distortion for short-term use will be more prevalent in the U.S.S.R., and it is a long-range weakness in Soviet science. There will always, of course, be a compromise in any country between short-term needs and long-term objectives, but the great danger of the planned state is that it will always over-emphasize the short-term objective. This is in some ways evident in the Soviet Union where problems of the moment have certainly led to the over-emphasis of physics as compared with chemistry and biology. Strangely enough, however, it does *not* seem to have led to an over-emphasis of applied science compared with pure science. In fact, in many ways the emphasis is the other way round compared with the U.S.A., but there seems at the moment to be a shift towards more emphasis on the applied side.

There have been many statistics and much talk about Soviet education and the rate of production of scientists and engineers, but there are many difficulties in comparing the East with the West. In the first place, "science" in the U.S.S.R. includes the humanities and social sciences and comparisons are often distorted on this account. Secondly, it is extraordinarily difficult to separate the training of technicians from that of scientists and engineers. The usual comparison overlooks altogether the on-the-job training of skilled workers in Britain and North America. Finally, there is a fundamental question of outlook. If in the U.S.S.R. almost all educated people are engineers and scientists, then most administrators in government and industry must come from this group. The comparisons of rates of production of engineers and scientists alone are therefore based on the tacit assumption that anyone taking a B.A. is of no further use to society. In fact, with our educational system administrators come mainly from this group, and if we assume that a general education is useful we should be comparing total university graduates, not just engineers and scientists. If we do this the figures are much more comforting.

There is, however, one very serious point which cannot be

overlooked. At the present time our so-called broad education is, in fact, totally lacking in any appreciation of science or its relation to society. Can people thus educated expect in the future, as in the past, to occupy commanding positions in society? I doubt it. I think, therefore, that the main problem in education is to rescue the humanities from a dangerous position by recognizing that they are an essential part of a broad education, but not the whole of it; and that people broadly educated in the real meaning of the term will be vitally needed in the future. If we do not recognize this we may find ourselves, willy-nilly, following the Russians in down-grading the importance of all education other than scientific. To my mind this would be the real disaster of the educational "race."

One other point which concerns education is the position of the universities in scientific research. In the U.S.S.R. the universities have been down-graded largely in favour of Institutes of the Academy of Sciences, and most graduate work in science, for example, is done in institutes rather than in universities. This means the separation to a considerable degree of teaching and research, and in my opinion has gone much too far. It emphasizes how necessary it is here to provide proper support for university research in order to avoid a similar situation.

It is when we come to the planning and the organization of science that the *paper* comparisons, charts and so forth, which compare the U.S.S.R. with the West are apt to deviate the most from reality. If the Iron Curtain is impenetrable enough and we sit for long enough on one side or the other we are apt to forget that those on the other side are still people. People everywhere will always be unco-operative towards planners, plans will always be partially evaded, and will always look infinitely better on paper than in practice.

The organization of everything in the U.S.S.R., including science, is much more monolithic than in the West, but it is totally wrong to assume, as is often done, that everything in the U.S.S.R. and nothing in the West is centrally planned. The differences between the East and the West are sharp and striking, but are by no means as great as might be imagined from an isolated and detached point of view.

In fact, without the slightest intention of being critical or insulting, one might suggest that the motives underlying GOSPLAN, the State Planning Commission of the U.S.S.R. are almost identical with those of the Treasury in Ottawa. The difference is merely that the Treasury works under much greater handicaps. I am sure that there are occasions when the Comptroller of the Treasury or the Secretary of the Treasury Board wish fervently that they could consign difficult people in organizations like the Research Council to Resolute Bay for the good of their souls and the greater glory of centralized control.

A second point is that the centralized planning of science in the U.S.S.R. is not nearly as complete or effective as it appears to be. From the point of view of the East-West struggle this is, of course, a pity because if their planning was perfectly successful everything would be so stagnant that we would have nothing to worry about. There are many reasons for the fact that planning in all countries always mercifully falls short of complete success. In the first place the motives of a man who is doing something are not always those of his boss. Hence people often succeed in doing the right thing when they are apparently following orders to do the wrong thing. Secondly, the planee always has more influence than appears on paper. Plans are made from below to a considerable degree, passed upstairs for approval, and back down again as orders. The result is that, even if science in the U.S.S.R. appears to be totally planned from above, many people are doing things they themselves thought of. This point was made to me most emphatically everywhere I went in the Soviet Union.

Finally, I think it is true that people with strong convictions will usually get around the rules in any country. Those with no convictions will always be planned by somebody, somehow, no matter what the system. Whatever the differences in the social systems, freedom of the individual, or standard of living, I am sure that the motives of the man in the laboratory are the same everywhere, and are compounded of a mixture of interest in the work and ambition to get ahead. The good man will always try hard to do the things he wants to do, and by and large will be fairly successful under any system. Circumventing rules may, however, be very time-consuming.

We in the West must watch ourselves to see that we do not lose our advantage by following the example of the U.S.S.R. in over-planning science. Let us remember that satellite launchings are not science. The U.S.S.R. has the great advantage of a docile population in planning large engineering and technological projects. We have an equally great advantage in the lack of a monolithic system which enables us to avoid over-planning science. Let us make sure that we do not lose our advantage simply because we do not understand the difference between science and technology.

Ultimately the thing that must be remembered is that the only level which really counts in science is the working level. There is a lot of paper and a lot of noise at the higher levels, and the noise is louder in the East than in the West, but none of this matters much. At the working level there really is not much difference between science in the U.S.S.R. and in the West.

INDUSTRIAL SCIENCE

Industrial Research

JUNE 14, 1956

. . . THE DEVELOPMENT of industrial research in Canada is following the general pattern of the development in the United States, but one war behind. Thus at the end of World War II Canada was left in a much more highly industrialized position and with much more industrial research going on: just as was the United States at the end of World War I. There is, however, one striking difference. During the last war, in Canada as in the United States, most research done was at government expense. The result is that Canada has never gone through a period where there is a large amount of industrial research in industry and a small amount in government institutions. Branch plants meant that most Canadian firms did no research at all at the beginning of the last war, and thus were not capable of undertaking research during the war. This largely prevented the placing of research contracts with industry by government as was done in the United States, because it is obvious that it is possible to place a research contract only in an organization which already has much experience in research and has a large and competent research staff. . . .

Why Do Research?

One important point is the question of the financial practicability of carrying out research in an industry, and the minimum size which an industry must be in order to afford a research laboratory. There is certainly a minimum size for a research laboratory, although there is some difference of opinion as to what this minimum size may be. There is I think a tendency, particularly in the United States these days, to make the minimum rather large. It certainly seems to be the thought of the Armed Forces that nothing can be accomplished for less than, say, $100,000,000. This overlooks the fact that prior to the war many first-rate research laboratories, both in industry and in universities, were relatively small, and I think that there is no question that an organization does not have to be big to carry out excellent work. However, it cannot be too small. It is true that there have been cases of a single individual in a small university carrying on work of first-rate quality. Such a man in a university has much more freedom than he would have in industry, however, and it is almost impossible to imagine a single individual doing anything worth while in industrial research. A fairly conservative estimate probably is that the minimum size of an industrial research laboratory is about 10 scientists and 10 technicians. This means a cost somewhere in the neighbourhood of $150,000 to $200,000 per year. Such an organization, provided that it is given freedom and that it is able to limit its objectives to a relatively small number of problems, can certainly produce worth-while results. . . .

Technical Knowledge

There are a number of ways in which technical knowledge can be obtained by a firm which is not itself doing research.

(a) The knowledge is obtained from its principals in the United States or United Kingdom. Basically this is the Canadian situation and is quite stable. Canada trains scientists, the scientists emigrate to the United States, they are hired by American firms, their salaries are paid from the money that is obtained from Canadian branches in exchange for the information resulting from their research. It is, of course, possible for this to go on forever and for Canada to continue as a scientific colony of the United States. The result, of course, is perfectly satisfactory provided that one is willing to accept, first, that nothing will ever be done for the first time by a Canadian firm, secondly, that we will in general be at least five years behind the United States in everything that we do, and, thirdly, that Canadian firms will always be at a considerable distance from their technical advice with all the attending complications.

It can, of course, be argued with complete justice that the value obtained from such an arrangement is enormous in proportion to the cost to the Canadian firm. I do not think that one can dispute the advantages of a technical association with a large American firm. What one can dispute is the fact that the payment to the American firm is made in cash rather than in kind. I can see no reason why a Canadian firm in these circumstances cannot spend the corresponding amount of money on research and then exchange results with its principals in the United States. This is a feasible arrangement provided that the Canadian firm with its relatively small laboratory concentrates on one or two problems and acts as the source of American information on them while it in return obtains American information on everything else. This kind of set-up is certainly possible on a defence and government level and would appear to be equally possible industrially. What is not possible is to have a relatively small staff and employ them on a multitude of problems. If this happens, sufficient weight is not put on anything to make the results worth while.

(b) One solution of the problem which is often suggested

and sometimes tried is to obtain an expert, possibly a high-priced one, and ask him to sit in the Canadian organization and keep track of everything that is done by the American counterpart. This is supposed to result in having all the information possessed by the Americans immediately available to the Canadians. There is one flaw, which is that no expert remains an expert when he is placed in the position of being merely a post office. In consequence, it is not possible to hire for this kind of work anyone who can do the job properly.

(c) Another solution is to employ consultants. In the first place, one can use the ordinary type of consulting firm, that is, consulting engineers, chemists, metallurgists, and so on. Within limits this is of considerable help. There is no doubt that for small firms who possess no technical staff a consulting firm is of great assistance. It is also true that on some very large projects a consulting engineer may become a specialist and work continually in one field. This is true, for example, in the design of bridges and things of this sort. In general, however, consulting firms have too broad a variety of problems to deal with and too much in the way of *ad hoc* problems to be a proper substitute for a research organization. Furthermore, even for trouble-shooting they are divorced from the individual firm to too great an extent to have an intimate familiarity with its problems. In some cases continuing and rather large contracts exist between an industrial firm and a consulting one. This may mean that the consulting firm maintains staff on a long-term basis for the sole purpose of dealing with the problems of one industrial firm. This essentially consists of setting up a research laboratory but locating it inside the quarters of a consulting firm. It is much better than merely using the consulting firm intermittently, but in general will probably be just as costly as setting up a research laboratory inside the company.

A second solution in the absence of a research laboratory is that of using university consultants. This is again only partially successful. It is, however, a successful extension of a company's research in the more basic fields. In general, all large American firms which run research laboratories of their own also use university consultants to a considerable extent. The comparatively small extent to which university staffs in chemistry and physics are used in Canada as consultants is probably merely an indication of the lack of research within industry itself.

(d) Another method of having research done outside is to make use of government laboratories. . . .

OCTOBER 30, 1956

. . . Actually, it is important to realize that it is neither possible nor desirable to expand Canadian industrial research too fast. What is needed is a continuing, steady, healthy growth, with emphasis on quality. Certainly bad research is worse than none. Also, it is essential that management do not go into research half-heartedly, but rather with the realization that research is not a luxury, but something which must be done to stay alive. There is, in fact, far too much of a tendency today to regard both research and scientific manpower as commodities which can be obtained by making provision for them in the budget.

In fact, there is a good deal of similarity between the development of research and the breeding of horses. If it were suddenly decided tomorrow that there was a vital need for ten million horses, the problem might be aided, but would certainly not be solved by budgeting for them, or by appointing committees or having conferences. The important thing would be to start breeding more horses as soon as possible, in order that they could in turn breed more horses, and so on. Similarly, the development of industrial research involves first the strengthening of university post-graduate schools, and, secondly, ensuring that the best of the graduates remain in the universities to train further research students. It is possible to have first-rate university research with little or no industrial research, and in fact this has been our history. It is absolutely impossible to have first-rate industrial research without first-rate university research. . . .

The Development of Industrial Research in Canada

OTTAWA BOARD OF TRADE ANNUAL DINNER
APRIL 6, 1961

SCIENTIFIC RESEARCH is much in the public eye these days. As a result there are two opposing tendencies: one is to exaggerate the accomplishments of research and to assume that scientists can fix everything up for everyone, overlooking economic and human troubles; the other is to decry the achievements of science, to attempt to set up a bogus rivalry with the humanities, and to make much of the "full man" even if he has an empty stomach.

I would like to discuss, as unemotionally as possible, the present Canadian situation. This is not easy to do. As is the case when dealing with economics, any suggestions which are made are apt to be quoted without qualifying clauses. Thus the simple statement that it would be nice to see more industrial research in Canada is apt to be regarded as critical of industry, as ultra-nationalistic, as anti-American, and as showing ignorance of economics.

In fact, a recent article in a prominent publication from Toronto has attempted to set up a "war" between the National Research Council and industry over research. This mythical antagonism was developed by the simple process of quoting statements (removed from their context) and having someone comment on them, and by a process of successive expurgation arriving at as controversial a result as possible. I would therefore like to draw attention to some grave defects in our position, but at the same time to emphasize the difficulty of correcting them. In doing this I will discuss some of the problems, considering in order research in universities, in government, and in industry. The reason for doing this is that it is possible to draw analogies about industrial research from the older and wider experience in Canada of university and government research. Both of these are in a much more highly developed state than is research in industry.

The Canadian situation presents some singular difficulties. Fifty years ago the United Kingdom and the United States were highly developed while Canada was still mainly a producer of raw materials. Industrial research on any scale is only about 50 years old in any country and at first it grew very slowly. In the last 20 years there has been an explosive growth, with many accompanying difficulties in leading scientific countries like the United States and the United Kingdom. In this same period, research in Canada has got going on an appreciable scale for the first time. As a result we have to catch up and keep up, both at the same time. This is a great difficulty and it is one which faces all newly-emergent countries which are making a quick transition to a high degree of industrialization.

It is instructive to look back at the history of Canadian universities, especially in science. Originally, of course, Canada was a colony in every sense of the word, and our educational pattern was appropriate for a colony. In the very early days Canadians went abroad for almost all their higher education, and it is only in the last 150 years or so that university education has existed at all in Canada. We first developed undergraduate education, but up to 40 years ago we were totally dependent on the United States, the United Kingdom, and Europe for all post-graduate education. Research in a Canadian university was a rare event in those days. After World War I there was a great development of graduate work in Canadian universities, but it was only after World War II that the Canadian universities really came into their own as centres of research and graduate education. This development is still going on, and still has a considerable distance to go. We are, however, well on our way towards competing with the best institutions in other countries in university science.

It is interesting to see what *might* have happened. It certainly would have been possible for us to have no university education in Canada at all, and to depend on other countries entirely. Similarly, it would have been possible to stick to the undergraduate or junior college level, and to depend on other countries for everything else. It would, however, have been an intolerable situation. We would have been electing for a permanent colonial status, for a permanent low educational level, for a permanent standard of living based on such a status, and for a permanent loss of our brightest people by emigration to more developed countries.

We owe a tremendous debt to the United States, to Britain, and to Europe for carrying us for all these years. I do not think, however, that we can in any way be accused of a narrow nationalism if we insist on proper universities of our own or show some pride in our university achievement. It is certainly not anti-American to feel that Canadian universities should be as good as those in the United States. Nor is it disrespectful to Harvard or Oxford or Cambridge to suggest that McGill and Toronto should be great centres of post-graduate work. In short, it is not anti-anything to feel that we should have a healthy Canadian science, Canadian art, Canadian literature, and so on.

As with the universities, so things have gone with government science. We *might* have evolved in Canada in such a way that we remained dependent on other countries for research aimed at developing our natural resources—agriculture, forestry, mines, and so forth. We might have depended entirely on our larger partners for research on defence problems, and for research by government aimed at secondary industry. In fact we did not. Because of specific Canadian problems, because of the disadvantages of research at a distance, because of economic factors, and perhaps because of national pride, Canadian government research developed early to a high state of quality in such fields as mining and agriculture. We now have facilities and people in these fields in government laboratories as good as (or better than) those anywhere else in the world. There is no doubt, also, that because these laboratories had a primary Canadian objective they have had a large share in the rapid development of our natural resources. Further, our *per capita* expenditure on research in these fields compares favourably with that in any other country. Again, we have grown up and passed the colonial stage.

The same is true of government research aimed at secondary industry such as is done in the National Research Council, in Atomic Energy of Canada Limited, and in similar fields. Here again we have not sat back and let others do things for us. Canadian government expenditure on civilian research is among the highest *per capita* anywhere in the world, and it has a high reputation. In defence research, the position is more difficult. Our total defence effort is not on as large a scale, proportionately, as that of the United States, and there are difficulties of which we are all aware in developing weapons for ourselves because of our smaller requirements. There are also difficulties in production-sharing. None the less, we have a major defence research effort, and one of the highest quality, in the Defence Research Board.

We have thus grown up as far as university and government research are concerned, but what I really want to discuss is the research which industry in Canada does for itself at its own expense. Research in industry in Canada is by no means in as advanced a state as that in universities and government, and it presents peculiar and difficult problems. It will, I think, be a help in approaching the situation if we consider two aspects of the question quite separately. First, is the present situation satisfactory? If not, can anything be done about it?

As far as the present situation is concerned there *is* research and development being done in Canada by purely Canadian companies and by Canadian subsidiaries. However, there are many Canadian firms, perhaps it would be fair to say most Canadian firms, which do no research or development. As a result the over-all performance of research and development by Canadian companies is relatively small. There is also another point which is of importance. In many cases when a Canadian company, particularly a subsidiary, does research it is almost all concentrated on things with a relatively short-term objective. There are comparatively few Canadian companies who do much of the really long-range type of applied research which will pay off in 10 or 20 years, rather than immediately. In other words, the more imaginative type of industrial research is often missing in Canada. (Please note the qualifying words, "often," "usually," and so on, in the above. I am not saying that no Canadian companies do research, that none do much, or that none is of high quality. What I *am* saying is that many do none, few do long-term work, and the over-all result is disappointing.) There is a trend which is decidedly in the right direction, but things are moving slowly, and in some quarters are at a stand-still.

The question I would like to ask is, "Can we face a *permanent* situation in which we are largely technically dependent on other countries?" If this is to be our situation can we ever expect to do anything better than it is done elsewhere? Technically we owe everything to the United States and the

United Kingdom, but technically we are still a colony. As far as research is concerned, industry is now largely where the universities were in 1930. Can we face a situation in which it remains there? It seems to me that we can answer only by saying that the situation is not permanently acceptable. If this is so, what can we do about it?

The first thing that should be emphasized, and that cannot be emphasized too much, is that there are many difficulties in doing research in a smaller country. However, the Swiss and the Swedes and the Dutch do it, so it cannot be impossible for us. There are two distinct problems: that of the purely Canadian company, and that of the subsidiary. The purely Canadian company is afflicted with competition from larger corporations to the south, with smaller markets, with tariff barriers against export, and many other problems. Nevertheless there are companies that have been conspicuously successful in performing research for themselves, one of them over a period of 40 years, and are now largely manufacturing things they themselves developed. But the going has been hard, and there are many companies which have not followed their example. In general, the pulp and paper and chemical industries are much ahead of the others. There have also been examples of initiative and self-sufficiency on the part of some of the smaller electrical firms, and of other companies. On the whole, one might say that the trend is right, but that the present situation is rather discouraging.

The real problem, however, is the Canadian subsidiary. Small companies cannot do much research and development, and the large companies are predominantly subsidiaries. Any major change in the situation must come about by means of an increase in the research performed in Canada by firms controlled from elsewhere. In considering this problem there are several things I am *not* going to suggest.

First, I will not enter into the general question of foreign control of Canadian industry. Far too many people seem to be emotionally involved in this at the moment, and in any case it is beyond my competence.

Second, I certainly do not suggest that any firm do research in Canada if it is uneconomic. Industrial research is a practical not an altruistic activity. It is not to be done for emotional reasons. However, it should not be avoided for emotional reasons, or because of inertia.

Third, Canadian subsidiaries are at present getting a great deal of technical information from their parents or affiliates elsewhere. Nothing must be done to interfere with this. Any technical assistance that can be obtained is obviously worth while. In fact we owe our degree of industrialization and our standard of living mainly to British and American technical competence and research.

Fourth, some subsidiaries do a lot of research in Canada. However, many do none and are technically completely dependent on their parent company. Others do a fair bit, but it is almost all along the lines of product development. Relatively there is not much long-term applied research of a rather fundamental kind. Nor, on the development side, is there enough high level development, rather than the trivial modification of items developed elsewhere. Again, one must recognize the problems of smaller markets. It is difficult to see how we can do well in exports if we never have anything to offer which is better or cheaper than that produced elsewhere. The Minister of Trade and Commerce has emphasized this. If, however, we never do research or development ourselves we never *will* have anything better or cheaper to offer.

Finally, it is perhaps worth pointing out that if we are to have any effect on the situation in 1965 or 1970 we must start now.

If it *is* assumed that a company can do research in Canada, and that it is not uneconomic, what are the advantages? The first advantage, of course, is that the company will normally go bankrupt if it does not do research. However, the real

Canadian problem is that subsidiary companies *can* do no research in Canada, but stay alive because of work done for them elsewhere. Even so, there are some distinct advantages to doing it on the spot rather than by remote control. First, it gives the whole organization (parent and subsidiary) a source of extra well-trained scientists. If there is a shortage of well-trained people in the United States and Britain, why not take advantage of the surplus of Canadians, but without asking them to leave Canada? Any company undertaking long-term applied research of high quality in Canada would have the inside track over the swarms of peripatetic foreign recruiting agents that now come through each spring. Secondly, the mere presence of a research laboratory will raise the level of technical competence of the company as a whole. It will also have the great advantage of bringing Canadian management into real contact with new developments at a much earlier stage. Thirdly, another factor which cannot be overlooked is that of size. Too often methods developed for a larger operation are merely transferred to Canada and scaled down. There may be many cases in which a fundamentally different approach would be more appropriate for our smaller operations or for our climatic conditions. Fourthly, in the long run, a worth-while research effort in Canadian companies would be bound to raise the standard of our whole technical effort. Finally, it might perhaps do something for our self respect. It would certainly do much for the morale of Canadian scientists and engineers.

It seems to me that there *is* one way in which Canadian subsidiaries could make a considerable increase in their research effort, which would be in no sense uneconomic, and which at the same time might help them to present a picture to the public of a really Canadian enterprise. At the present time, as research laboratories in the United States grow, there is a considerable tendency to decentralize research activities to some degree to branches and subsidiaries in the United States.

There are, in fact, many advantages to such decentralization, the workings of Parkinson's law being one of them. There is no doubt that many such laboratories have passed the optimum size. Is there any reason why the effort in certain fields, long-term as well as short, should not be decentralized to the Canadian subsidiary, rather than to the Oklahoma branch?

The difficulty at present is that when the Canadian subsidiary does research at all the effort is apt to be spread thin and tends to cover everything in a shallow and *ad hoc* way. Could not the subsidiary carry on the main effort in one field for the whole organization, and be dependent mainly on the parent in other fields? In this way the magnitude of the research effort of the subsidiary could be comparable to its relative size, and it could be a first-rate effort in depth. Certainly there are in many cases advantages in decentralization because it leads to more initiative and imagination. Certainly, also, it cannot be argued that research in Canada is more expensive than that in the United States. If such decentralization were carried out the advantages of raising the technical level of the company and having an effect on the thinking of management would be obtained as a bonus.

To sum up, it seems to me that no matter what the difficulties may be, the present situation is intolerable as a continuing one. There are grave difficulties in doing much about it, but the decentralization to Canada of research, especially long-range research, is in no way impossible. We cannot do anything sudden or dramatic, but perhaps we can do something. At least we should give all possible encouragement to industrial research in Canada, and perhaps even put on a little pressure. The trend is encouraging; perhaps we could speed things up. At least in industrial research we should remember that far from being an important country we are rather a newly-emergent underdeveloped nation.

INTERNATIONAL SCIENCE

The International Unions

INTERNATIONAL UNION OF GEODESY AND GEOPHYSICS DINNER
TORONTO, SEPTEMBER 3, 1957

I WOULD LIKE to add one more expression of pleasure to those which have gone before that the International Union of Geodesy and Geophysics is meeting in Canada, and that we are thus hosts to so many scientists distinguished in these fields. It is also a special honour and pleasure that the Union should have chosen to come to Canada at a time when the International Geophysical Year is in progress, and when geophysics is thus very much in the news.

Geophysicists are accustomed to speak of various regions in Canada. To the ordinary Canadian, however, there are just two such regions: Toronto and elsewhere. I am sure President Smith [Sidney Earle Smith, President, University of Toronto, 1945–57] will pardon me if I express the hope that in addition to meeting here and enjoying the hospitality of Toronto you will also try to see something of that other Canada which lies beyond the boundaries of the city of Toronto.

Owing to the extent and the geological diversity of Canada, the number of geophysical problems to be investigated is very great. Also, the location of the North Magnetic Pole within our boundaries, and the proximity of the geographical pole pose special problems. The relatively small population of Canada magnifies the formidable character of these tasks, and though the geophysical profession here is active, independent, and very much inclined to stand on its own feet, we cannot help but be conscious of the desirability of scientific aid from abroad, at least in the form of moral support and advice on problems of an international nature.

The meetings here are also advantageous to us from another point of view. We are a new country, rather far away from the older centres of scientific work, and just beginning to come to the front in science ourselves. In the past we have suffered greatly from isolation. Modern means of communication are diminishing this isolation, and are making it possible for us to be hosts to many scientific visitors, singly or in groups such as yours. It is a great honour to us and a great advantage to scientific work in this country to be hosts today to such a group as your International Union. Your presence here cannot fail to be a stimulus to Canadian work in your field.

It seems to me that there are certain problems facing science in the next few years, and that International Unions can make a contribution to the solution of some of them. I would, therefore, like to discuss a few aspects of science from this point of view. One of the things which is obviously of concern to the Unions is the question of internationalism in science. Over the past few centuries science has reached a position of internationalism which is shared by very few other human activities.

This position was not achieved suddenly or without difficulty. In fact in the Middle Ages there was little of an international nature about science. Most princelings kept a tame alchemist or two about the premises, but they certainly did not encourage the dissemination of scientific information, and in fact were inclined to use the axe to discourage it. Slowly, fostered by the newly formed scientific academies, information began to get about, and a surprisingly free flow developed. . . .

Internationalism in science developed over the years to a remarkable degree. The free interchange of ideas, free publication, and so on were a remarkable tribute to the universal nature of science, and until recently there has been little sign of nationalism in science. Of course, no situation is ever perfect and the author indexes of books, perhaps, permitted a prediction of the nationality of the author. (He mentioned himself the most often, followed by his compatriots, but this was merely normal human frailty.)

We are, of course, in a difficult period internationally at present, and it becomes a major task of scientific unions to uphold the internationalism of science, and in fact to hold up science as an example in this respect which might be followed in other fields of human endeavour.

There is, however, a rather worrying ambiguity about what is meant by internationalism in science. The original idea of Unions was, surely, a place where scientists could get together co-operatively, irrespective of nationality. It was certainly not an attempt to organize science on a federal basis among states. In other words, scientific internationalism originally meant total lack of nationalism, and not internationalism in the diplomatic sense. The trend has, however, been away from these ideas. Since the war the Unions have been organized under the I.C.S.U. which receives support from UNESCO. As a result the Unions seem to me to be tending to be less international in the more healthy sense of the word, and becoming more concerned with national affairs rather than less.

The Unions themselves realize this, and there is a considerable degree of ambivalence in their operations. Thus most Unions hold meetings of the Union and a Congress simultaneously, the former being afflicted with national delegates and the latter being free from national considerations. Similarly most Unions have a peculiar organization whereby the commissions operate more or less free from the national bias of the other operations. It does seem to me, however, that it would be a great pity if in the future the Scientific Unions were to become more and more concerned with national questions, and I for one hope that there may be a movement away from the "federal" outlook, back to the ignoring of nationalism.

A problem which is closely linked to that of nationalism in science is that of communication. The whole scientific effort is obviously dependent on the free transmission of knowledge. There are today two essentially opposite facets to this problem. In the first place there are those, and I am not among them, who are worried at the great increase in scientific publication and the increasing difficulty of keeping up with what is going on. The worriers along these lines seem to feel that the whole method of disseminating scientific information which was built up over the centuries—meetings, papers, abstracts, monographs, and so forth—is breaking down. Figures are cited to show the impossible amount of reading which must be done, the number of pages published, and so on. What seems to be forgotten is how much of this is trivial or positively worthless. After all, no one is foolish enough to want to read all the trade journals published in the world. It seems to me that it is not much more difficult to keep up than it used to be: all that is necessary is to be willing to read and to be able to ignore the trivial. What is essential, however, is that abstract journals should not be allowed to languish. In many fields there is a real need for international co-operation in abstracting, and a real place for the activities of scientific unions in this field.

The main thing seems to me to avoid hysteria and in particular to avoid the suggestions of the information officers and the data-processors. The more extreme members of this school seem to envisage an automatized scientific literature which is just as horrifying as other automatic devices. An assortment of machines, punch-cards, and information specialists will make sure that everyone gets what he should read. Carried to its logical conclusion this system would ensure that no one would ever read anything who understands it. It seems to me that there is really very little to worry about, provided that those who specialize in the organization of scientific literature are kept firmly in their place.

A much greater problem of communication, however, is that of secrecy. Secrecy of one sort or another, industrial or military, has always been with us and has always been objectionable. However, until relatively recently the information thus concealed was so trivial that it made only a negligible difference to the progress of science.

The recent expansion in industrial and military research has made a considerable change in the picture. Most industrial research is on the development side and from a scientific point of view is still trivial, but there is a sufficient amount of work of importance to make the future of industrial secrecy a matter of some concern. On the military side the situation is similar. Most military research is of considerable technological but only of trivial scientific importance. However, there are certain fields, and geophysics is one of them, where the body of military information is of respectable dimensions.

It is obvious that, in the long run, secrecy means nationalism of the most extreme form. In fact, one could visualize a situation in which a field of physics in one country could be almost unrecognizable to workers in another. However, this is not likely to occur. All that is apt to happen is a tremendous waste of effort, and a certain delay. Here again, however, it is essential that everything possible should be done to break through this type of narrow nationalism. There is no doubt that meetings like the present one, and co-operative efforts like the I.G.Y. can do much to restore the normal means of scientific communication.

There will, of course, always be problems of technological as opposed to scientific secrecy. However, technological information is only of short-term value as opposed to scientific information, and technological secrecy can do relatively little harm.

At all events the meeting of people and exchange of ideas are the surest ways of breaking down artificial barriers, and this and other Unions have a continuing task to perform along these lines.

There is another set of problems confronting us which stem from the fact that science has become useful in the last 30 years or so. In earlier days the tangible results of the applications of science were not great, and the public were not impressed by science except as an intellectual exercise and attainment. Today, however, science gets the credit not only for what it does, but for many things it does not do, and is cited as responsible for everything from tooth-paste to atomic power. The result is obvious. As long as science was merely an intellectual exercise society was willing to provide a small amount of support and no interference. Once the game becomes regarded as of practical importance, however, support goes up and so does interest and interference. Naturally enough, society cannot remain aloof from something which affects it in a vital way.

There are, however, some unfortunate aspects to this interest on the part of society. There is in the first place a lack of understanding of the difference between science and technology. Secondly, science has become a "spectator-sport" with the box office and publicity looming large. Scientists, of course, are human and rather like publicity, but it is not entirely an advantage in the academic pursuit of knowledge.

Another outcome of the situation is that because society is footing the bill and is interested in the results there is naturally an attempt going on to make research "efficient." The problem is that those who are trying to make science efficient do not know what it is, and are trying to apply the normal criteria of cost accounting. This inevitably leads to enthusiastic attempts to "plan" science, and I think that such planning is probably the greatest danger facing science today. A thoroughly planned, well-co-ordinated effort will drive ahead relentlessly to the obvious conclusion, but is most unlikely to produce a new idea. Teams are all very well, but neither a team nor a committee has ever been known to have an idea, and ideas come from people, not teams. . . .

Science and International Affairs

MACDONALD COLLEGE
NOVEMBER 4, 1960

IN THE PAST, with one exception to which I will return, science has been singularly free from government control or interest. In fact, science has developed almost altogether apart from the influence of society. There have been exceptions: Galileo had to say the earth was flat; bishops denounced science as the work of Satan; pulpits thundered on the subject of evolution; and today satellites and fall-out are treated with emotion rather than reason. By and large, however, science has ignored such attacks and has developed on its own. In fact, it has been able to do so to a much greater extent than the humanities, which have always been in danger of suppression and distortion by theologians and ideologists. . . .

The one exception to which I referred in which the freedom of science was infringed was in the case of the alchemists. They were hired to transmute base metals into gold and all that they did was kept secret. It is curious that the only pronounced case of secrecy in modern times has been in atomic energy, which again involves transmutation. The reason, of course, is not hard to find. The ability to make gold or to make atomic bombs gives power to him who possesses it, and once power is involved science cannot be left in isolation. As a result two phenomena are occurring which are changing the picture of things: one is the impact of science on government; the other is the impact of government on science. Both will be fraught with difficulties in the next few decades.

At any rate, whatever the causes and the problems, science is now of decided interest to governments because of its effect on the economy, because of its defence implications, and because it is impinging more and more on things international and political (in the wider sense). In short, national prestige is now largely a question of technological achievement, which in turn is largely a matter of science. This has meant two things: first, governments can no longer afford to stay out of scientific research (in Canada the federal government now spends 200 million dollars a year on it) and secondly, governments are more and more in need of advice on scientific matters, and cannot afford to depend on casual and occasional advice from outside.

The involvement of the government in science raises two separate problems. The first is how to operate successfully a research laboratory under the bureaucratic and centralized methods of government operation. This is a major problem; a still more difficult, and in fact almost impossible problem is to operate a laboratory under international rather than national bureaucracy. The second problem, and the one to which I propose to devote my attention, is the question of scientific advice and the interplay of scientific and political questions. This is a serious problem, but it is so new that not much thought has yet been given to its wider implications.

It can, I think, be argued that scientists have not been very helpful. Scientists in general, and scientific societies in particular, have argued vigorously that scientists must be consulted on every phase of international affairs which has any technical content. They have not, however, often realized that in many cases the technical aspect may be small, and the political aspect large, and that in international affairs it may often be necessary for valid reasons to ignore competent scientific advice. It must not be forgotten that, if science has an impact on political considerations in the international sphere, then political considerations must also have an impact on scientific policy. There must be mutual understanding between the scientist and the diplomat, and suggestions by vocal groups such as those who edit the *Bulletin of the Atomic Scientists* that all would be well with the world if it were only run by nuclear physicists hardly need rebuttal. It is an unfortunate fact that no matter how objective a scientist may be in his own work, he is just as emotional and illogical as anyone else once he gets outside it. Anyone who has ever attended a university faculty meeting will agree that neither a scientist nor a humanist is *ex-officio* easy to get along with.

On the other side of the picture, however, is the fact that no one has been more successful in international affairs than has the scientist. With an objective outlook in his own field, and able to meet those in other countries on common ground, the scientist has established an international framework which can serve as a model for all other fields of endeavour. It has not always been so. Until three or four hundred years ago science, like most things, was plagued by a combination of superstition, tradition, secrecy, and theology, and this together with poor communications led to the development of local schools of thought. Since then, however, until relatively recently, improved communications and relative freedom from economics and ideology have led to an almost complete absence of nationalism in science. . . .

There is no question that economic, military, and political considerations make it impossible for science to remain aloof as in the past. What we must do is to adjust our thinking in such a way as to play a proper part in national affairs without losing the non-nationalism in pure science which has been achieved over the centuries.

In dealing with international organizations which concern themselves with science to a greater or lesser degree it is important to distinguish between governmental and non-governmental organizations. As two extreme cases consider the Faraday Society and the United Nations. The Faraday Society is a United Kingdom physical chemistry society. It stages symposia and publishes a journal. Its membership includes people from every country in the world, and it is rare for one of its symposia to have fewer than ten nationalities represented. Although located in the British Isles it has met abroad, and has even had the temerity to elect a Canadian president. There is no doubt about its international character: there is also no doubt that its members belong as individuals and do not in any sense represent the government of the country they come from. As a result it is totally non-governmental in character, and would never, and could never, concern itself with government policy in Britain or elsewhere. At the opposite pole is the United Nations. It is purely governmental, and makes no sense in any other context. When it concerns itself with questions with some scientific content, its members are still speaking on behalf of their governments, and political considerations will always be expected to outweigh scientific ones. Any scientist who gives assistance must realize that he is merely an adviser, and that government policy will always be the determining factor.

In between these two extreme cases there is a variety of organizations, concerned to a greater or lesser degree with science, which possess a partial governmental character. UNESCO is inter-governmental in nature but, because of its advisory committees, acts as somewhat less than a purely governmental body. I.C.S.U., the International Council of Scientific Unions, obtains subventions from UNESCO and from National Academies, but has a mixed function as a governmental body and one composed of individual scientists. The International Unions, further down the scale, are financed nationally, and tend to have some degree of rotational and regional representation on committees, but are almost non-national in character: at least, to a considerable degree people speak for themselves rather than for nations.

In addition there are organizations of limited blocks of nations with common objectives, or of nations in limited geographical regions which have formal organizations partially concerned with science. Examples are the nations concerned with the Colombo plan, with economic co-operation in Europe, with the development of Africa south of the Sahara, with NATO, and with the British Commonwealth. The result is a plethora of meetings which results in turn, as one prominent American scientist puts it, in all scientists over 50 spending far more time talking about their work than doing it. On the other hand, such organizations have a real and a legitimate interest in science, and somehow the best possible advice must be channelled to them. Life can, however, get very complicated when virtually all these organizations are interested in the same subject, for example, space research. Certainly the only possible solution is close co-operation between scientists and those concerned with foreign affairs. The advice of scientists is essential to prevent diplomats from making foolish decisions where science is involved, but it is no solution to let scientists take over and make foolish decisions on international politics, as seems to be the idea of some rather vocal pressure groups.

For example, obviously the views of scientists on the technical side of disarmament are of the greatest importance, but the decisions are ones which involve political negotiation and are not basically scientific at all. There is certainly no excuse for the view that nuclear physicists have special qualifications to negotiate on disarmament, or that the views of nuclear physicists or physical chemists or microbiologists as a group should be taken any more seriously than those of any other group with a sincere interest in world peace. It should not be forgotten that it is possible for an individual to be very brilliant in one field, scientific or otherwise, and a crackpot in another in which he has limited knowledge.

All this leads, I am convinced, to a fundamental schizophrenia affecting science at the moment: the question of nationalism *versus* non-nationalism in outlook. The days of aloofness are gone, and this must be recognized, but the integrity of science must not be lost in the process of such recognition. There is no question that the outlook of science today must be nationalistic in certain aspects. The first of these is in financing. The day of the private benefactor of universities and institutes is largely over, and today governments at one level or another are the main support of universities and of science. To a considerable extent, *via* the development contract, this is true even of scientific work carried on in industry. It is impossible, therefore, for national considerations to be overlooked, and the structure of science in any given country, or its importance relative to the humanities, must largely be determined or warped by government policy regarding the priority of certain fields. An example of this is the relatively much greater emphasis on physics than chemistry in the U.S.S.R. as compared with the United States.

A corollary to the importance of national financing is the importance of recognition of science at the government level. Countries vary widely in their recognition of the importance of science. In trying to achieve such recognition publicity comes in, and it is impossible to avoid considering the status of Canadian physics or of Norwegian biology. Allied with this is a justifiable pride in accomplishment, or shame at the lack of it. It is certainly not reprehensible to point out that

Canadian medicine has a high reputation, or that Canadian agricultural research has a long-continued tradition of high quality. All this, however, does tend to suggest to the public mind that science is local rather than world-wide, and to obscure the almost complete freedom of exchange of results and ideas on a personal basis which ignores all boundaries and iron curtains.

Again, science cannot be ignored as part of national economic development; it is the inevitable penalty for being useful. Applied science is certainly a definite factor in the health of the national economy, and because pure science is the foundation of applied science it also cannot be ignored in relation to economic policy. Again, there are possible dangers in the over- or under-emphasis of certain fields. In a similar way, no one can suggest that the applications of science to defence can be regarded from any point of view but a national one. Finally, when scientists are acting as advisers on government policy, they are certainly tied in with the nation's affairs. In all this, however, if science is to preserve its integrity, and also its usefulness, there are many nationalistic pitfalls which must be avoided.

One problem today is the prevalence of false publicity and false claims for priority of discovery. This is not new (in fact almost every town in central Europe has two things in common: first the former residence of one of Goethe's mistresses; and secondly the home of the discovery of wireless telegraphy). However, such claims—I mean of the latter type—have been getting more frequent and more nationalistic in recent years, and the fault is by no means all on the eastern side of the Iron Curtain. The main thing, I think, is to avoid the outlook of certain small magazines which insist that nothing elsewhere (including science) is as good as it is at home, and that through science all is for the best in this best of all possible worlds.

Scientific publications are not a source of major trouble at the moment. Previous difficulties of access to Soviet journals are mostly over, and the free circulation of knowledge is again going on. Security restrictions are becoming limited to more trivial things, and exchanges of people are helping to straighten the situation out. One major difficulty, however, looms on the horizon. An increasing nationalism is destroying the old custom of publishing only in one of the three or four main languages. More and more papers of some real value are beginning to appear in obscure languages, giving rise to translation difficulties, and the major build-up of Chinese science will make these acute within a decade.

All these are minor troubles. There are, however, two major problems which result from the impact of national and international questions on science. These involve the questions of objectivity and of prestige, and I am afraid that there is no simple solution to either of them.

Scientific objectivity has in the past been something of which all could be proud, and in fact might be taken as an indication that similar objectivity in more difficult fields was not impossible to attain. Today, however, there are many disturbing problems, and the controversy over bomb tests and fall-out has done great damage to the public picture of the scientist as an objective person. Two of the major exponents of diametrically opposite points of view have been a distinguished chemist and a distinguished physicist, neither of whom is expert in genetics. In my mind there is no doubt that both have distorted the scientific facts on the basis of emotional, and sincere, views on disarmament. This raises the difficult question of how the scientist is to keep his scientific judgment clear of all bias, but at the same time, as a person, be free to express his own political views whatever they may be. The difficulty is whether the scientist can speak as an individual without his scientific prestige slopping over into the political realm. Certainly if a scientist is sufficiently distinguished it is difficult for his views on non-scientific matters not to carry excess weight on the basis of his scientific reputation. The difficulty becomes almost insuperable if the question involved has a partial technical content. As science gets more and more bound up with national and international matters this question will become steadily more difficult, and there is a danger that in the process scientific thought will be carried back to the atmosphere of emotion and prejudice of the Dark Ages.

The question of prestige is also very troublesome. It is most unfortunate to regard science as a race with anyone, and the atmosphere of the Olympic Games with national statistics on gold medals should be avoided if at all possible. However, there is no doubt that today questions of prestige are of great importance in international affairs. There is thus every justification for the distorting of priorities for reasons of prestige, and there is far more behind satellite launchings than science and the spirit of discovery and adventure. It is this area which produces the maximum clash between scientific and political questions, and such a clash cannot be avoided in the kind of world in which we live. The morals of science are thus being slowly broken down until they are not much better than those of the market-place, but I doubt if there is much we can do about it.

Apart from these questions of the interplay of science and politics there is, I think, one special place where scientists can make a contribution to international good will. This is by maintaining personal contacts with people of many nationalities. Workers in science have a great advantage in this connection in that they start on common ground, they already know foreign workers from their publications and from correspondence, and they get together with reasonable frequency by means of international meetings or exchanges. It is thus possible to meet political opponents in a more friendly way than can any other group. There is no question that the presence of international graduate schools such as have existed at Oxford, Cambridge, and McGill for many years can be a great stimulus to international understanding. The same is true of the National Research Council where at any time we have people of about 20 nationalities and where we have over the last ten years had people from virtually every country in the world. The continuation of this type of internationalism is, I think, of major importance.

Summing up, I think it can be said that the changing position of science in its relation to society is producing many problems. It is most important that we avoid weakening the structure of science by a narrow nationalism. At the same time it is equally important that scientists do their part by trying to keep their advice objective even when political considerations may, in the last analysis, be paramount. To do this will require in the future considerably more tolerance and mutual respect from scientists, diplomats, and politicians than has been evident in the past; and it must be admitted that scientists have by no means been above reproach. The problems in the future will certainly be no less complex than they are at the moment, and scientists must be prepared to make the maximum possible contribution to world affairs. They can do this, however, only by an appreciation of the other factors, and the other people, involved; certainly they cannot make such a contribution by insisting, as they do far too often today, that they occupy a position of special privilege.

APPENDIX

Profile: The New President of I.C.S.U.

New Scientist XII (October 5, 1961), p. 32

AMONG ITS STAFF in Ottawa, it is a common remark that the National Research Council has been fortunate not only in its four presidents but even more so in the order in which they have held that office. This remark is particularly apposite to Dr. E. W. R. Steacie, the current President, who last week was elected President of the International Council of Scientific Unions. Steacie, who by preference and by accomplishment is predominantly the academic research scientist, became President of N.R.C. at a time when the Council had consolidated the tremendous expansion of the war years and had disposed of two major responsibilities that had absorbed much of its energies—atomic energy and research for the military services.

The problems facing the Council when Steacie became president were to strengthen its own laboratories, to rededicate its research programme to projects more appropriate for its long-range responsibilities, and to continue to expand its support for research in the universities. For these tasks Steacie was eminently qualified not only by training and inclination but also by his ability as an administrator, his sure feeling for scientific values, and his imaginative grasp of scientific organization.

For a pure research scientist, Steacie's early academic career was not exactly normal. Born in 1900, he first attended the Royal

Military College at Kingston, Ontario, and then studied chemical engineering at McGill University. After taking a B.Sc., however, he was persuaded to do graduate work in physical chemistry and coming under the influence of Dr. Otto Maass quickly became absorbed in problems of basic research. Having achieved his Ph.D. in 1926, he continued as a lecturer at McGill and, shortly after, started that research in chemical kinetics which has remained his major scientific interest. During 1934–35 Steacie took a leave of absence from McGill to spend a year in Europe. During this period he worked with Professor K. F. Bonhoeffer in Frankfurt and with Professor A. J. Allmand at King's College, London. Returning to McGill he remained there as Associate Professor of Chemistry until 1939 when he joined the National Research Council as Director of the Division of Chemistry.

With his appearance at N.R.C., Steacie not only continued and expanded his systematic and inspired researches on photochemistry but also demonstrated that the qualities that had already brought him recognition as a research chemist were more than sufficient to make him one of the most efficient administrators in the government service. In his early days at the Council, Steacie acquired a reputation for his outspoken criticism of government red tape. At one stage it was rumoured that he was so fed up that he was ready to return to university life, convinced that good research was impossible in a government laboratory. Fortunately for N.R.C. and for Canadian science he did not make this move but remained at the Council, where his insight and his administrative talents helped to prove how false his own fears were.

For a period during the war Steacie was second-in-command to Sir John Cockcroft at the British-Canadian Atomic Energy Project in Montreal, which at that time was administered by N.R.C. It is probable that, in those hectic days, surrounded by a team of *prima donna* scientists and charged with responsibility for a "crash" scientific programme, Steacie learned much about what to avoid in scientific administration and consolidated his own ideas on scientific organization. Certainly he came out of this experience with strong views on the evils of secrecy and with a firm conviction that it was not wise to allow nuclear scientists to assume that they spoke for the totality of science.

Back at the Research Council after the war, Steacie resumed his researches and quickly re-established his laboratory as one of the leading centres of chemical kinetics in the world. The first edition of his monograph, "Atomic and Free Radical Reactions," appeared in 1946 and quickly became the standard text on the subject. In 1948 he was elected into the Royal Society.

It was during the period immediately after the war that Steacie conceived the imaginative plan that developed into the N.R.C. post-doctorate fellowships. Through this scheme, N.R.C. has been able to bring to Canada for short definite periods outstanding young scientists from all the advanced countries of the world, and by this means has been able to avoid two of the major dangers of national laboratories—a narrowness of outlook arising from lack of contact with the academic world, and inflexibility of staff caused by fixed establishments.

In 1950 Steacie became Vice-President (Scientific) of the National Research Council, a position which gave him wider responsibilities within the Council and, also, made him formally a member of the Honorary Advisory Council for Scientific and Industrial Research—the Research Council proper. In 1952, on the retirement of Dr. C. J. Mackenzie, Steacie became President, that is he became chief executive officer of the laboratory system comprising N.R.C. as well as the presiding officer of its Council, a body whose primary interest and direct responsibility was the support and promotion of research in the universities.

Steacie's success as a fundamental research scientist and his wider activities in the scientific societies made him a natural leader in shaping policy and developing plans for academic support in Canada. Even his most enthusiastic supporters, however, have been surprised at the tact and skill with which the hitherto blunt and outspoken scientist has led the Council, has moulded its opinions, and has adroitly steered it around all the dangers that are inherent in an academic advisory body. As an executive officer Steacie has been equally effective; his systematic methods, his immediate grasp of the essentials of a situation, and his lack of any element of indecision have allowed him to fulfil these responsibilities with ease and efficiency and yet with time for consideration of the wider issues. Although as a scientist, Steacie always denounced vehemently red tape and all other forms of bureaucratic control, as an administrator he quickly established cordial relations with other government departments and a feeling of mutual respect with Treasury Board officials. He argued frequently and strongly with these officials but always on points essential to his concept of sound scientific policy and never on the inconsequential procedures dear to the hearts of bureaucrats.

As President of the Research Council, Steacie has been forced to take an increasing interest in the relation of science to government or, as he once described it in a lecture, "the impact of government on science." By inverting the emphasis, he wished to point out that, because science has now become of great importance to governments as a major expenditure and as an asset that cannot be neglected, scientists can no longer expect to be left alone in their ivory towers but must reconcile themselves to increased attention from the political authorities. Although he realizes the danger of government interference with science, distorting its structure and pushing it in wrong directions, Steacie recognizes the inevitability of the process and has always been prepared to accept the ascendancy of political considerations as long as the integrity of the scientific process has not been compromised. He has had scant sympathy for those scientists who by their success in research have been led to think that all political problems can be solved readily if only the scientists are in control.

More and more Steacie has been forced by his position to participate directly in international scientific activities. Since he became President of the National Research Council and chief scientific adviser to the Canadian government he has found himself increasingly called upon to represent Canada on inter-governmental scientific committees and at meetings of international scientific organizations. Not only does the National Research Council have the responsibility in Canada for the International Council of Scientific Unions but, in addition, Steacie is the obvious choice to represent Canada on such bodies as the NATO Science Committee, the Advisory Committee for Natural Sciences Programme of UNESCO, and the British Commonwealth Scientific Committee.

In observing the operation of these international bodies Steacie has characteristically separated the trivial and superficial from the truly scientific and essential. As one who has faced and overcome the difficulties of the national administration of scientific laboratories he has been less than enthusiastic about schemes for international institutes or foundations. He possesses too keen a sense of the problems of scientific organization to wish to replace the known methods of national governments by the hazards of international bureaucracy and inter-governmental committees.

Although he would defend to the last the position of science as a supranational activity transcending national borders and would support strongly the necessity for international standardization through the Unions and the co-ordination of national programmes of research through some mechanism such as the I.G.Y., he believes that unless there are overriding considerations the responsibility for scientific programmes can best be handled at the national level where administrative procedures have been developed and where financial control ultimately resides.

With this experience behind him, Steacie is now accepting the responsibilities of the presidency of the I.C.S.U. at a time when that organization is approaching a critical period in its existence. In the I.G.Y., the I.C.S.U. demonstrated with extraordinary success a pattern for international co-operation by which large and extensive national programmes of research were co-ordinated without over-all international control or a large international secretariat. The control and organization of the I.G.Y. remained in the hands of the scientific organizations with only a minimum of political interference. If the I.C.S.U. is to maintain its position at the apex of international scientific organizations it must demonstrate not only that it can maintain the impetus of the I.G.Y. but also that its structure and methods are adequate for the problems of international co-operation that must be faced in the future. This is the sort of challenge that Steacie will face with enthusiasm and with a rare combination of administrative skill and scientific insight.

Although his current activities leave him little time for matters unrelated to science, Steacie could in no sense be classified as the narrow scientist. In his younger days he was a proficient golfer but lately he has forgone this pleasure as he found that to play golf well required more time than he could spare. Until very recently, he maintained a great interest in skiing and each week-end during the winter months was to be seen on the trails in the Gatineau Hills near Ottawa. It was the practice in his laboratory for the post-doctorate fellows to take one afternoon each week to go skiing with "the boss" whenever the weather was fine and the snow good. Of late years, however, Steacie's outside interests have been concentrated on the summer cottage which he himself built on a lake in the hills north of Ottawa and where he spends his summer holidays and all possible week-ends.

Steacie has no use for sham or affectation and at one time was known for his impulsive Irish temper. Although he has mellowed much with age and responsibility, he is still not one to suffer fools gladly particularly when an administrative official introduces red tape unnecessarily or some misguided scientist advocates some woolly idea on scientific organization. Where the good of science or the efficiency of N.R.C. is concerned, Steacic has no patience for incompetence but in his personal relations with his staff he is the most considerate of men.

J. D. Babbitt

Randall Library – UNCW
Q127.C2 S7 NXWW
Steacie / Science in Canada; selections from the s
304900219142Z